Mastering ArcGIS Server Development with JavaScript

Transform maps and raw data into full-fledged web mapping applications using the power of the ArcGIS JavaScript API and JavaScript libraries

Ken Doman

PUBLISHING

BIRMINGHAM - MUMBAI

Mastering ArcGIS Server Development with JavaScript

First published: September 2015

Production reference: 1110915

Published by Packt Publishing Ltd.
Livery Place
35 Livery Street
Birmingham B3 2PB, UK.

ISBN 978-1-78439-645-9

www.packtpub.com

Credits

Author
Ken Doman

Reviewers
Rusty Green
Matthew King
Dr. Jason E. VanHorn
Dr. Ashwini Wakchaure

Commissioning Editor
Pramila Balan

Acquisition Editor
Vinay Argekar

Content Development Editor
Rohit Kumar Singh

Technical Editor
Saurabh Malhotra

Copy Editors
Janbal Dharmaraj
Kevin McGowan

Project Coordinator
Mary Alex

Proofreader
Safis Editing

Indexer
Hemangini Bari

Production Coordinator
Shantanu N. Zagade

Cover Work
Shantanu N. Zagade

About the Author

Ken Doman has worked in the geographic information system field since 2007. In his first GIS job, he was given a stack of maps, some AutoCAD data, and a shoe box full of address notes and was told to make a web-accessible GIS application out of them. Since then, he has strived to make geographic data available and easily accessible over the Internet. He is an ESRI certified web application developer and has created many public web applications using ArcGIS software.

Currently, Ken works for Bruce Harris & Associates, Inc., a mapping and custom development company that specializes in land records and ESRI software solutions. There, he creates custom web, desktop, and mobile applications using JavaScript, .NET, and ArcGIS software.

He has also been a book reviewer for *Building Web and Mobile ArcGIS Server Applications with JavaScript*, *Packt Publishing* by Eric Pimpler and *ArcGIS for Desktop Cookbook*, *Packt Publishing* by Daniela Cristiana Docan. You can learn more about Ken at http://raykendo.com.

I would like to first thank my wife, Luann, who put up with many of my late night book-writing sessions. She supported me and pushed me when I needed it. I would also like to thank the city of Jacksonville, Texas, the city of Plantation, Florida, and Bruce Harris & Associates, Inc. for giving me the opportunities to learn the lessons that I was able to pour into this book. Finally, I would like to thank god, the source from whom all blessings flow.

About the Reviewers

Rusty Green is an enthusiastic software developer who specializes in creating simple and reusable web technology solutions and enjoys collaborating on open source projects. He has a diverse background in technology and holds an MS in information systems engineering and management, a BS in geospatial information systems, an AS in architectural technology, and an array of various certifications. Rusty has held various roles throughout his career, such as a senior software developer, GIS analyst, corporate faculty member, and director of technology and software engineering.

You can learn more about him at rustygreen.com.

Matthew King is a software development professional from Brandon, Florida. He holds a bachelor's degree in information technology from the University of South Florida. He has a wide range of development experience, including web, desktop, mobile, database, and GIS applications.

Currently, he works in the City of Tampa as an applications systems analyst and database administrator. He spends his free time on mobile, website, and game development as well as actively participating in development user groups throughout the Tampa and Orlando area. Matthew is the co-organizer of the Tampa Mobile Applications Developer user group. He is also a veteran of the United States Navy with 22 years of service.

He enjoys blogging about various technologies that interest him on his website at http://mattkingit.com. He is now pursuing a Microsoft Certified Solutions Developer certification in Web Applications and Windows Store Apps.

Dr. Jason E. VanHorn is an associate professor of geography at Calvin College and an expert in cartography, Geographic Information Systems (GIS), and remote sensing. He holds a PhD from The Ohio State University, a master of science degree from Texas A&M University, and a bachelor of arts degree from Indiana University. He is the chair of the GEO department at Calvin College and his current research foci involve 2D and 3D visualization of physical dune geomorphology and security issues dealing with terrorism and Homeland Security analysis. He has authored numerous publications and online GIS applications and consults on a variety of national and international projects.

Dr. Ashwini Wakchaure, originally an architect and planner from India, has a PhD in design, construction, and planning from the University of Florida. She has been working with GIS technologies for over 10 years now. Primarily an ESRI GIS user, she has extensively worked with the .Net technologies involving ArcObjects, Python, and more recently, in the web GIS development area.

www.PacktPub.com

Support files, eBooks, discount offers, and more

For support files and downloads related to your book, please visit www.PacktPub.com.

Did you know that Packt offers eBook versions of every book published, with PDF and ePub files available? You can upgrade to the eBook version at www.PacktPub.com and as a print book customer, you are entitled to a discount on the eBook copy. Get in touch with us at service@packtpub.com for more details.

At www.PacktPub.com, you can also read a collection of free technical articles, sign up for a range of free newsletters and receive exclusive discounts and offers on Packt books and eBooks.

https://www2.packtpub.com/books/subscription/packtlib

Do you need instant solutions to your IT questions? PacktLib is Packt's online digital book library. Here, you can search, access, and read Packt's entire library of books.

Why subscribe?

- Fully searchable across every book published by Packt
- Copy and paste, print, and bookmark content
- On demand and accessible via a web browser

Free access for Packt account holders

If you have an account with Packt at www.PacktPub.com, you can use this to access PacktLib today and view 9 entirely free books. Simply use your login credentials for immediate access.

Table of Contents

Preface

In this digital age, we expect maps to be available wherever we are. We search for driving directions on our phone. We look for business locations and nearby restaurants using apps. Maintenance personnel use digital maps to locate assets several feet underground. Government officials and C-level executives need up-to-date location data to make important decisions that affect many lives.

ESRI has been an industry leader for 30 years in the realm of digital mapping. Through products such as ArcGIS Server, they empower companies, government agencies, and digital cartographers to publish maps and geographic data on the Web. Websites and applications written with the ArcGIS API for JavaScript make those maps available on both desktop and mobile browsers. The API also provides the building blocks needed to create powerful applications for data presentation, collection, and analysis.

While there are plenty of examples, blog posts, and books to get you started in developing apps with the ArcGIS API for JavaScript, they don't often go that deep. They don't discuss the pitfalls, limitations, and best practices for developing applications with their tools. That next step is what this book tries to accomplish.

What this book covers

Chapter 1, Your First Mapping Application, introduces you to ArcGIS Server and the ArcGIS API for JavaScript. In this chapter, you will learn the basics of creating a map and adding layers.

Chapter 2, Digging into the API, gives an overview of the many API components, widgets, and features that are available within the ArcGIS API for JavaScript. This chapter offers explanations of how these components can be used.

Chapter 3, *The Dojo Widget System*, explores the Dojo framework, the comprehensive JavaScript framework the ArcGIS API for JavaScript is built in. This chapter explores the inner workings of Asynchronous Module Design (AMD) and how to build custom widgets for our mapping applications.

Chapter 4, *Finding Peace in REST*, examines the ArcGIS REST API, which defines how ArcGIS Server communicates with the application in the browser.

Chapter 5, *Editing Map Data*, covers how to edit geographic data stored in ArcGIS Server. This chapter examines the modules, widgets, and user controls provided by the ArcGIS JavaScript API.

Chapter 6, *Charting Your Progress*, looks at how you can convey map feature information through charts and graphs. The chapter not only discusses charts and graphs made with the Dojo Framework, but also looks at how to integrate other charting libraries, such as D3.js and HighCharts.js.

Chapter 7, *Plays Well with Others*, discusses ways to integrate other popular libraries into an application written with the ArcGIS API for JavaScript. This chapter also explores combining the framework with jQuery, Backbone.js, Angular.js, and Knockout.js.

Chapter 8, *Styling Your Map*, covers the use of CSS styling in the mapping application. This chapter examines how the Dojo framework lays out elements, and explores adding the Bootstrap framework to style the mapping application.

Chapter 9, *Mobile Development*, looks at the need to develop with mobile in mind and discusses the pitfalls of attempting to make your app "mobile-friendly". This chapter implements the dojox/mobile framework, provided by the Dojo framework, as a way to style elements for mobile use.

Chapter 10, *Testing*, discusses the need for test-driven and behavior-driven development. This chapter discusses how to test applications using the Intern and Jasmine testing frameworks.

Chapter 11, *The Future of ArcGIS Development*, looks at new mapping and application services such as ArcGIS Online and Web AppBuilder. You will learn how to develop an application around an ArcGIS Online webmap.

What you need for this book

For all the chapters in the book, you will need a modern browser and a text editor to create files. Whichever text editor you choose is up to you, though one with syntax highlighting and some sort of JavaScript lining would help. For exercises in *Chapters 5*, *Editing Map Data* and *Chapter 9*, *Mobile Development*, you will also need access to a web hosting server to test files written in Java, PHP, or .NET. Finally, for the exercise in *Chapter 10*, *Testing*, you will need an up-to-date version of Node.js installed locally on your machine.

Who this book is for

This book is for web developers with some experience with ArcGIS Server, or geospatial professionals with some experience writing HTML, JavaScript, and CSS. This book assumes that you have viewed some of the ArcGIS API for JavaScript examples provided by ESRI. More importantly, this book is for someone who wishes to dive deeper into application development using ArcGIS Server. Other books provided by Packt Publishing, such as *Building Web and Mobile ArcGIS Server Applications with JavaScript* by Eric Pimpler or *Building Web Applications with ArcGIS* by Hussein Nasser, may be more suitable for an introduction to ArcGIS Server and the ArcGIS API for JavaScript.

Conventions

In this book, you will find a number of text styles that distinguish between different kinds of information. Here are some examples of these styles and an explanation of their meaning.

Code words in text, database table names, folder names, filenames, file extensions, pathnames, dummy URLs, user input, and Twitter handles are shown as follows: "We're working with census data, let's call it `census.html`."

A block of code is set as follows:

```
<!DOCTYPE html>
<html>
<head></head>
<body></body>
</html>
```

When we wish to draw your attention to a particular part of a code block, the relevant lines or items are set in bold:

```
<!DOCTYPE html>
<html>
<head>
  <meta http-equiv="Content-Type" content="text/html;charset=utf-8"/>
  <meta http-equiv="X-UA-Compatible" content="IE=Edge" />
  <meta name="viewport" content="initial-scale=1,
  maximum-scale=1,user-scalable=no"/>
  <title>Census Map</title>
  <link rel="stylesheet" href="http://js.arcgis.com/3.13/
  esri/css/esri.css" />
  <style>
    html, body {
      border: 0;
      margin: 0;
      padding: 0;
      height: 100%;
      width: 100%;
    }
  </style>
  <script type="text/javascript">
    dojoConfig = {parseOnLoad: true, debug: true};
  </script>
  <script type="text/javascript" src="http://js.arcgis.com/3.13/"
  ></script>
</head>
```

New terms and **important words** are shown in bold. Words that you see on the screen, for example, in menus or dialog boxes, appear in the text like this: "We positioned it precisely in the top right, and left a little gap for the **Census** button to be centered vertically."

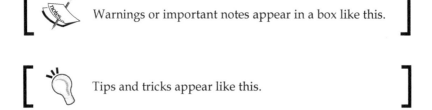

Warnings or important notes appear in a box like this.

Tips and tricks appear like this.

Reader feedback

Feedback from our readers is always welcome. Let us know what you think about this book — what you liked or disliked. Reader feedback is important for us as it helps us develop titles that you will really get the most out of.

To send us general feedback, simply e-mail feedback@packtpub.com, and mention the book's title in the subject of your message.

If there is a topic that you have expertise in and you are interested in either writing or contributing to a book, see our author guide at www.packtpub.com/authors.

Customer support

Now that you are the proud owner of a Packt book, we have a number of things to help you to get the most from your purchase.

Downloading the example code

You can download the example code files from your account at http://www. packtpub.com for all the Packt Publishing books you have purchased. If you purchased this book elsewhere, you can visit http://www.packtpub.com/support and register to have the files e-mailed directly to you.

Downloading the color images of this book

We also provide you with a PDF file that has color images of the screenshots/ diagrams used in this book. The color images will help you better understand the changes in the output. You can download this file from http://www.packtpub.com/ sites/default/files/downloads/6459OT_ColorImages.pdf.

Errata

Although we have taken every care to ensure the accuracy of our content, mistakes do happen. If you find a mistake in one of our books — maybe a mistake in the text or the code — we would be grateful if you could report this to us. By doing so, you can save other readers from frustration and help us improve subsequent versions of this book. If you find any errata, please report them by visiting http://www.packtpub. com/submit-errata, selecting your book, clicking on the **Errata Submission Form** link, and entering the details of your errata. Once your errata are verified, your submission will be accepted and the errata will be uploaded to our website or added to any list of existing errata under the Errata section of that title.

To view the previously submitted errata, go to `https://www.packtpub.com/books/content/support` and enter the name of the book in the search field. The required information will appear under the **Errata** section.

Piracy

Piracy of copyrighted material on the Internet is an ongoing problem across all media. At Packt, we take the protection of our copyright and licenses very seriously. If you come across any illegal copies of our works in any form on the Internet, please provide us with the location address or website name immediately so that we can pursue a remedy.

Please contact us at `copyright@packtpub.com` with a link to the suspected pirated material.

We appreciate your help in protecting our authors and our ability to bring you valuable content.

Questions

If you have a problem with any aspect of this book, you can contact us at `questions@packtpub.com`, and we will do our best to address the problem.

1
Your First Mapping Application

Let's say you have a map. You've digitized it using ArcGIS Desktop, a sort of desktop mapping and analysis software provided by ESRI. You've gone through the painstaking process of plotting points, connecting lines, and checking the boundaries of polygons. You've added nice background aerial imagery, and you've applied all the parts that make a map readable.

How are you going to share this map with the general public? You could post it in the public information office, but citizens have complained about that location being too remote, and down too many flights of stairs underground. You could make a thousand printed copies, but that would be terribly expensive.

If you have the **ArcGIS Server** software running and connected to a web server, you can publish your map online, and serve it through a website running with the **ArcGIS JavaScript API** (http://developers.arcgis.com/javascript).

The ArcGIS JavaScript API is a JavaScript library that works with ArcGIS Server to connect the map maker with the general public. The map maker can use an ESRI product, such as ArcMap, to generate a map document. That map maker can then publish the map document through ArcGIS Server. From there, a web page that has been loaded with the ArcGIS JavaScript API can draw the map in the browser, and let the general public pan, identify, and interact with the map.

In this chapter, we'll be covering the following topics:

- The requirements for creating a web mapping application using ArcGIS Server and the ArcGIS API for JavaScript

- The HTML head and body content necessary to serve maps with the JavaScript API

- How to create a map with the ArcGIS JavaScript API and add content

- How to make a map interactive

Features of the API

The ArcGIS JavaScript API provides many of the tools necessary to build robust web map applications. It can generate **slippy** maps, interactive maps that let the user pan and zoom in. The behavior is similar to Google or Bing Maps, but with your data. You're in control of the content, from background imagery to markers and popup content. With ArcGIS Server, you have control over how the maps are laid out, and which colors, styles, and fonts you use. The API also comes with custom elements that let you do everything from drawing on the map, searching for data, measuring things on the map, and printing the map in multiple formats.

The ArcGIS JavaScript API is built on top of the **Dojo** framework (www.dojotoolkit.org).You also have access to an extensive package of free HTML form elements, controls, and layout elements for your web applications because Dojo comes packaged with the API. These Dojo user controls have been tested in multiple browsers, and include an entire library of items that can be used to make a mobile application. While the ArcGIS JavaScript API is built with Dojo, it also works well with other libraries such as jQuery and AngularJS.

The ArcGIS JavaScript API was designed and built, along with an ArcGIS API, for Flash and Silverlight. Unlike other APIs which require specialized compilers, plugins, and related software, the ArcGIS JavaScript API can be written with a simple text editor and viewed on most common browsers without any special plugins. Since mobile browsers, such as Safari for iPad and Chrome for Android, do not support third party plugins, the ArcGIS JavaScript API is the preferred choice for creating interactive map websites for the mobile platform.

Now, it is possible to code a website using Windows Notepad, just like it's possible to hike up Mount Everest without a guide. But when things go wrong, you will probably want to use a free text editor with syntax highlighting and other features, such as NotePad++ (`http://notepad-plus-plus.org/`), Aptana Studio 3 (`http://www.aptana.com/products/studio3.html`), or Visual Studio Code (`http://code.visualstudio.com`) for Windows, Brackets (`http://brackets.io`) or Textmate (`http://macromates.com/`) for Mac, or Kate (`http://kate-editor.org/`), Emacs (`http://www.gnu.org/software/emacs/`), or vim (`http://www.vim.org/`) for Linux. If you want text editors that aren't free, but offer more features and support, you can check out Sublime Text (`http://www.sublimetext.com/`) or Webstorm (`http://www.jetbrains.com/webstorm/`).

The ArcGIS JavaScript API community

The ArcGIS JavaScript API has an active community of developers who are willing to help you along the way. ESRI has blogs where they post updates, and host meetups in various cities across the country and around the globe. Many ArcGIS JavaScript developers, both inside and outside of ESRI, are active on Twitter and other social media outlets.

You can find many resources to help you learn about the ArcGIS JavaScript API through books and the web. For books, you can check out *Building Web and Mobile ArcGIS Server Applications with JavaScript* by Eric Pimpler, *Building Web Applications with ArcGIS* by Hussein Nasser, and *ArcGIS Web Development* by Rene Rubalcava. For online resources, you can visit ESRI GeoNet (`https://geonet.esri.com/community/developers/content`), view the arcgis-javascript-api tag on GIS StackExchange (`http://gis.stackexchange.com/questions/tagged/arcgis-javascript-api`), or visit the ESRI GitHub page (`https://github.com/esri`).

Our first Web Map

Now that the introductions are out of the way, we should begin working with the API. In this chapter, we're going to look at some code that will make a simple, interactive map. The example will be a single-page application, with all the styling and coding on the same page. In the real world, we would want to separate those into separate files. For this example, this is what the project is comprised of:

- Setting up an HTML5 web page
- Adding the necessary styling and the ArcGIS JavaScript library
- Framing out our HTML content
- Setting up a script to create a map
- Loading a layer file
- Adding a click event that collects data from the map service
- Displaying that data on the map

Our assignment

We've just been asked by the Y2K historical society to make an interactive map application of the United States around the year 2000. They want the application to show the U.S. demographics including gender, age, and ethnicity, during that year. After reviewing the request from the client, we determined that the 2000 census data would provide all the mapping and demographics data we were looking for.

After a bit of research, we found an ArcGIS Server map service that serves the 2000 census data. We can use the ArcGIS JavaScript API to show that data on an HTML document. The user will be able to click on the map, and the application will display census data by state, census tract, and census block group.

Setting up the HTML document

Let's open our favorite text editor and create an HTML document. Since we're working with census data, let's call it `census.html`. We'll start with an HTML5 template. Browsers will recognize it as HTML5 by the appropriate document type at the top of the page. Our HTML5 page starts out as follows:

```
<!DOCTYPE html>
<html>
<head></head>
<body></body>
</html>
```

Starting from the head

The head of an HTML document contains information about the page including the title, metadata about the page content, **Cascading Style Sheet** (**CSS**) links to tell the browser how to render the output, and any scripts that the developer needs the browser to run before it reads the rest of the page. Here is an example of a simple webpage:

```
<!DOCTYPE html>
<html>
<head>
    <meta http-equiv="Content-Type" content="text/html;charset=utf-8"/>
    <meta http-equiv="X-UA-Compatible" content="IE=Edge" />
    <meta name="viewport" content="initial-scale=1,
    maximum-scale=1,user-scalable=no"/>
    <title>Census Map</title>
    <link rel="stylesheet" href="http://js.arcgis.com/3.13/
    esri/css/esri.css" />
    <style>
      html, body {
        border: 0;
        margin: 0;
        padding: 0;
        height: 100%;
        width: 100%;
      }
    </style>
    <script type="text/javascript">
      dojoConfig = {parseOnLoad: true, debug: true};
    </script>
    <script type="text/javascript" src="http://js.arcgis.com/3.13/"
    ></script>
</head>
    ...
```

Downloading the example code

You can download the example code files from your account at http://www.packtpub.com for all the Packt Publishing books you have purchased. If you purchased this book elsewhere, you can visit http://www.packtpub.com/support and register to have the files e-mailed directly to you.

Let's take a look at each of the items separately.

Meta tags and title tags

The title and meta tags in the head of the document tell browsers and search engines more about the page. See the following code for an example:

```
<meta http-equiv="Content-Type" content="text/html;charset=utf-8"/>
  <meta http-equiv="X-UA-Compatible" content="IE=Edge" />
  <meta name="viewport" content="initial-scale=1,
  maximum-scale=1,user-scalable=no"/>
  <title>Census Map</title>
```

Some tags tell search engines how to read and categorize the content of your website. Others, like the meta tags in this file, tell the browser how to display and manipulate the page. In the preceding code the first meta tag, we're establishing the character set used to render text. The second meta tag tells Internet Explorer browsers to load the page using the latest version available. The third meta tag tells mobile browsers that the content is scaled to the correct size, disabling the ability to pinch or spread your fingers on the screen in order to zoom the text in and out. This is different from zooming the map scale in and out, and this tag is required in most mobile browsers to zoom in and out of the map.

Cascading style sheets

The look of a website is determined by the CSS. These style sheets tell the browser how to lay elements on the page, what colors to make each element, how to space them out, and so on. You can see how they are arranged in the current document:

```
<link rel="stylesheet" href="http://js.arcgis.com/3.13/
  esri/css/esri.css" />
  <style>
    html, body {
      border: 0;
      margin: 0;
      padding: 0;
      height: 100%;
      width: 100%;
    }
  </style>
```

Here are three ways to organize and control the styling of elements by using CSS:

- First, we can apply styles to individual elements on the page with inline styling, by adding a style attribute to the HTML element (`<div style="...">`</div>` for instance).

- Second, we can apply styling to the whole page by using an internal style sheet, denoted by the `<style></style>` tags.

- Third, we can apply styles to multiple sheets by referring to an external style sheet, denoted by `<link rel="stylesheet" ... />`. For our single page application, we'll use both our own internal style sheet, and an external one provided by ESRI.

The ArcGIS JavaScript library requires its own style sheet to properly position the maps on the screen. Without this file, the map will not render correctly, and thus show the expected results. To load the required style sheet for the library, add a `link` tag and set the `href` attribute to point to the `esri.css` file.

If you're using version 3.2 or later, and your map appears on the page in a checkerboard pattern with the map tiles showing on every other square, the most likely problem is that the `esri.css` style sheet did not load. Make sure you reference the correct `esri.css` style sheet for the library version you're using. The following image shows an example of this behavior:

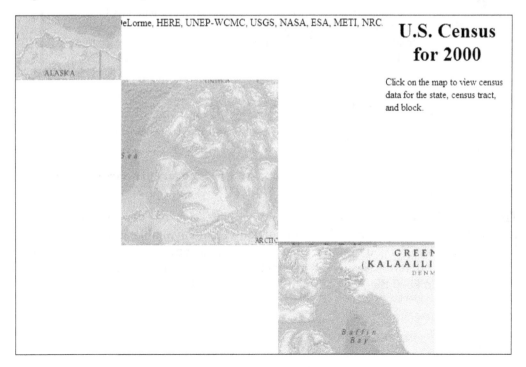

The Dojo configuration script

Our JavaScript code adds a variable that tells Dojo how it should load, in the first script tags. From this script, we can tell the browser how to interpret specially defined HTML elements in our document, whether we want the browser to cache all files, and even how to load packages and other libraries into the Dojo build system:

```
<script type="text/javascript">
  dojoConfig = {
    parseOnLoad: true,
    cacheBust: true
  };
</script>
```

In this case, we're telling Dojo to parse any specially decorated HTML elements in the body, and replace them with the appropriate Dojo widgets as the page is loaded. Using the `cacheBust` parameter, we are also asking the browser to use a timestamp when it loads the files, so that the files aren't cached in the browser. Adding timestamps forces the browser to load a fresh copy of the JavaScript file, rather than relying on a cached copy. Cached scripts under active development may not show the most recent changes you made, slowing development and increasing troubleshooting time.

 The script that loads the `dojoConfig` object must come before you load the ArcGIS JavaScript API. If the `dojoConfig` object is created after the API script reference, the `dojoConfig` contents will be ignored.

The ArcGIS JavaScript API script

The ArcGIS JavaScript library is the main tool you'll use to render, manipulate, and interact with geographic data from the ArcGIS Server:

```
<script type="text/javascript" src="http://js.arcgis.com/3.13/"
  ></script>
```

This application, as well as other applications in the book, use version 3.13 of the ArcGIS JavaScript API. It was the most current version available at the time the book was written. As you maintain these applications, be aware of version number updates. ESRI often releases new versions to add new features, fix bugs in previous versions, and to keep the API compliant with the latest browsers.

Moving from the head to the body

With our HTML head set up, we can focus on the body of the application. We'll add HTML elements to the body where the map and other information should go. We'll style those features from an inline style sheet. Finally, we'll write a script to handle the map creation, census data retrieval, and reacting to map events.

Framing the HTML body

Our client specified that they would like the app to show two panels, a main map panel, and a separate panel that explains what the user is supposed to do. We're going to fulfill the request by blocking off the sections using HTML div elements. The div elements are generic blocks of content in HTML. In the first div, we'll add a styling class of instructions, and fill it with the appropriate instructions. In the second div, we'll apply the specific element id of map, to tell ourselves and the ArcGIS JavaScript API where to put the map:

```
<body>
  <div class="instructions">
    <h1>U.S. Census for 2000</h1>
    <p>Click on the map to view census data for the state, census
    tract, and block.</p>
  </div>
  <div id="map"></div>
</body>
```

Adding a little style

We need to add some style to the new elements we added. To do this, we'll modify the original internal style sheet in the head portion of the application. Our client wants the map to take up the whole screen, with a little space in the upper right-hand corner for the map title and the instructions. The client hasn't decided on colors yet, but they have requested the rounded corners that everybody's putting on their websites today.

So, after reviewing the requirements, and looking up how to style the elements, let's add the following within the `<style></style>` element. The changes have been highlighted to help you see what has changed in the following code snippet:

```
<style>
  html, body, #map {
  border: 0;
  margin: 0;
  padding: 0;
  width: 100%;
```

```
    height: 100%;
  }
  .instructions {
      position: absolute;
      top: 0;
      right: 0;
      width: 25%;
      height: auto;
      z-index: 100;
      border-radius: 0 0 0 8px;
      background: white;
      padding: 0 5px;
  }
  h1 {
      text-align: center;
      margin: 4px 0;
  }
</style>
```

Here's an explanation of the style we've added. We want the HTML document, the map `<div>` to have no margin, border, or padding, and take up the full height of the page. We also want the `<div>` elements with the instructions class to be precisely positioned in the top right corner, taking up twenty-five percent of the page's width, and then its height will be determined automatically. The instructions block will be floating 100 z-index units towards the user (putting it in front of our map), and its bottom-left corner will have an 8 pixel radius curve in the lower left corner. It will have a white background, and a little padding on the right and left side. Finally, the title `<h1>` will be centered horizontally, with a little padding above and below it.

Adding a script at the end

If we look at our web page now, we won't see much. Just a title and instructions in the upper right-hand corner. We need to turn this plain page into a fully powered map application. To do that, we'll need to instruct the browser, by using JavaScript, on how to transform our map `<div>` into a map.

Just before the end of our body tag, we'll add a script tag where we'll write our JavaScript code. We put our script near the end of our HTML document because, unlike images and style sheets, browsers load script tags one at a time. While the browser is loading the script, it doesn't load anything else. If you put it at the beginning of the page, the user might notice a bit of latency, or a delay before the page loads. When we put a script at the end, the user is too distracted by the images and other elements on the page to notice when your script loads. This gives it the appearance of loading faster:

```
    <div id="map"></div>
    <script type="text/javascript"></script>
</body>
```

So, why didn't we load the ArcGIS JavaScript library at the end of the page? There will be times, especially if you are using other parts of Dojo, when we'll need the library to manipulate the page as it loads in the browser. In that case, we put the library reference at the head of the HTML document.

Now that we have a script tag to write some JavaScript, we can code up an interactive map. But, before we start writing code, we need to understand how to use the ArcGIS JavaScript API and Dojo. We'll start with a quick history lesson.

Back in the good old days of web development, and even continuing today, JavaScript libraries tried to avoid colliding into one another by creating a single global object (such as JQuery's $ or Yahoo's YUI). All of the library's features would be built into that object. If you've used the Google Maps API, you've probably used `google.maps.Map()` to create a map, and `google.maps.LatLng()` to load a point on that map. Each subpart of the library is separated by a dot (`.`).

Older versions of the ArcGIS JavaScript library were no different. All of ESRI's mapping libraries were loaded into the main "esri" object. You could create a map using `esri.map`, and load it with an `esri.layer.FeatureLayer` to show some data, for instance. The Dojo framework was similar, using the `dojo`, `dijit`, and `dojox` global objects.

But this tree-like approach to JavaScript library design has its downsides. As the library expands and matures, it builds up a lot of parts that developers didn't always use. We might use a library for one or two specific features, but we may not use every single tool and function the library offers. The parts of the library we don't use waste client bandwidth, bloat the memory, and make our app appear slower to load.

Asynchronous Module Definition

Both the ArcGIS JavaScript API and Dojo decided to handle the bloated library crisis by incorporating the concept of **Asynchronous Module Definition** (**AMD**). In AMD, a library is broken down into modular components. The developer can pick and choose which parts of library they want to include in the application. By loading only the parts we need, we reduce download times, free the browser memory of unused functionality, and improve performance.

Another advantage of AMD is name collision avoidance or the names of the variables where the libraries load are controlled by the developer. Also, the scope of the loaded libraries is limited to within the calling function, much like a self-executing statement.

In an AMD based application, we make a list of the library modules we want to use, usually in an array of strings that the library knows how to interpret. We then follow it up with a function that loads most or all of those modules into JavaScript objects. We can then use those modules within the function to get the results we want.

Loading required modules

To take advantage of the Dojo's AMD style, we're going to use Dojo's `require` function. In older examples, we would create multiple `dojo.require("")` statements that would load the parts of the ArcGIS JavaScript library we needed (and hopefully by the time we wanted to use them). But with AMD style, we use a single `require` function that requests the list of libraries we ask for, and loads them within a function that runs after all the libraries are loaded in the browser:

```
<div id="map"></div>
<script type="text/javascript">
  require([], function () {});
</script>
</body>
```

The `require` function takes two arguments, an array of strings that corresponds to folder locations in our library, and a function that runs after those libraries load. In the second function, we add arguments (variables within the parentheses after functions) that correspond to the libraries loaded in the list.

So, for this application, we'll need a few modules from the ArcGIS JavaScript API. We'll need to create an empty map and add data to the map in a form we call layers. We'll need to identify things on the map, retrieve the census data we need from the place we clicked on the map, and then display it:

```
<div id="map"></div>
<script type="text/javascript">
  require([
    "esri/map",
    "esri/layers/ArcGISDynamicMapServiceLayer",
    "esri/tasks/IdentifyParameters",
    "esri/tasks/IdentifyTask",
    "esri/InfoTemplate",
    "dojo/_base/array",
    "dojo/domReady!"
  ], function (
    Map, ArcGISDynamicMapServiceLayer,
    IdentifyParameters, IdentifyTask, InfoTemplate,
    arrayUtils
  ) {
    // code goes here
  });
</script>
```

Notice the order of the libraries loaded in the require statement, and the arguments in the following function. The first item in the list corresponds to the first argument in the function. It is a common error when creating more complex applications to mix up the order of the elements, especially if it has been revised multiple times. Make sure the items correspond as you go down the list.

You may have noticed that, while there are seven libraries loaded, there are only six arguments. The last library that loaded, the `dojo/domReady!` library, tells the require statement's second function not to run until all the HTML elements have loaded and are rendered in the browser.

The map object

Now that we have the components, we need to create an interactive map; let's put them together. We'll start by constructing a map object, which provides the basis and interaction platform that we'll need to work with. The map constructor takes two arguments. The first parameter, either the HTML node or the id string of a node, indicates where we want to put our map. The second parameter of the map constructor is an options object, where we add the configurable options that make our map work as expected:

```
function (
    Map, ArcGISDynamicMapServiceLayer,
    IdentifyParameters, IdentifyTask, InfoTemplate,
    arrayUtils
) {
  var map = new Map("map", {
    basemap: "national-geographic",
    center: [-95, 45],
    zoom: 3
  });
});
```

In the preceding code, we're creating a map in the div element with an id of map. In the map, we're adding a **basemap**, or a background reference map, in the style of National Geographic. We're centering the map at 45°N and 95°W, at a zoom level of three. We'll go into greater depth concerning these configurations in later chapters. If you're using a desktop browser to view the results, you should see the United States, including Alaska, and Hawaii, as in the following image:

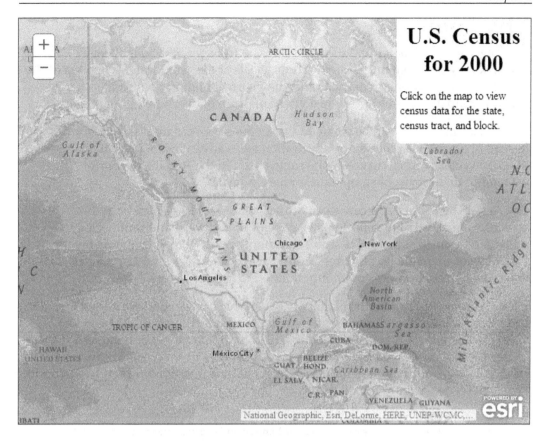

If you have experience working with Dojo, most element constructors have the options first, followed by the node or the id of the node. This is the reverse of how we construct a map. Remember, order is important.

The layers

Years ago, mapping departments drew maps on transparent Mylar sheets. Painstaking work went into drawing those sheets at the same scale. When the Mylar sheets were stacked on top of each other, and corresponding points on each layer were lined up, they would provide a visual mashup of overlapping map layers.

Today, the same effect can be created with a browser-based map application. Instead of clear Mylar sheets, the application takes advantage of layering image files and vector graphics to create the same effect. In the ArcGIS JavaScript API, we refer to these stackable map data sources as **layers**.

The ArcGIS JavaScript API can accept multiple types of layer files from different sources. Some layers are very flexible, and can be realigned and reprojected to line up with other map sources. These are commonly referred to as **dynamic layers**. Other layers are made up of images drawn at specific scales, and are not designed to be so flexible. These are commonly referred to as **tiled layers**.

In our current application, the National Geographic background we added is considered a tiled layer. Its pre-rendered content loads quickly in the browser, which is one of the advantages of using a tiled layer. On the other hand, our data source for the census data is a dynamic layer provided by an ArcGIS Server map service. Its dynamic nature helps it to stretch and line up with our tiled background. Here's the code we'll use to add the layer:

```
var censusUrl =
  "http://sampleserver6.arcgisonline.com/arcgis/rest/services/
  Census/MapServer/";
var map = new Map("map", {
    basemap: "national-geographic",
    center: [-95, 45],
    zoom: 3
  });

var layer = new ArcGISDynamicMapServiceLayer(censusUrl);

map.addLayer(layer);
```

So, if we take a look at the page at this point, we should see a map of the world with black lines surrounding each state, census tract, and block. We can zoom in to see more detail in the map. Your map should look something like the following image:

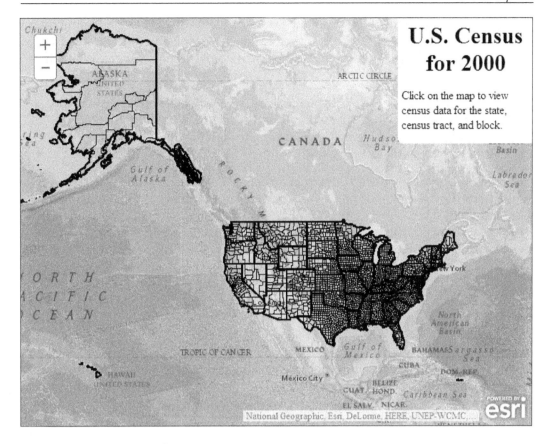

Adding some action

At this point, we have a map that draws all the state and census block and track data. We need more than that. We need a way for the user to interact with the site. The ArcGIS JavaScript API incorporates many tools provided by the native JavaScript language, plus new tools provided by Dojo.

Events

When many of us learned traditional programming, we learned that programs followed a linear design. They started on the first line and ended on the last line. The computer performed each computation in an ordered fashion, and would not go on to the next line until the current line had finished.

But JavaScript adds something different. JavaScript adds an event loop that monitors for specific website interactions, such as a change in a text blank. We can attach a function, commonly called an **event listener**, to a known element event. When that event is triggered, the event listener runs.

For example, buttons have a click event, and if we attach an event listener to the button's click event, that function will respond every time that button is clicked. We can also get information about the button through the data passed through the click event.

For our application, we want the map to do something when we click on it. If we look up the list of supported map events, we can attach an event listener using the .on() method. The .on() method takes two parameters, a string description of the event, and the event listener function we want to be called when the event occurs. A list of supported events can be found in the ArcGIS JavaScript API documentation:

```
map.addLayer(layer);
function onMapClick (event) {
}
map.on("click", onMapClick);
```

Before we can do anything with the map, we need to know if it has loaded yet. If we're running an updated browser on our fastest computer with a high-speed internet connection, the map may load immediately, but if we're testing from a mobile browser on an older smartphone, with less than stellar internet download speeds, the map may not load very quickly.

Before we assign the click event, we're going to test whether the map has loaded. If so, we'll add the onMapClick function as an event listener to the map's click event. If not, we'll wait until the map fires its load event to set the click event. Just so we don't have to repeat ourselves when adding the map click event, we'll enclose that assignment in another function:

```
function onMapClick(event) {
}
function onMapLoad() {
  map.on("click", onMapClick);
}
if (map.loaded) {
  onMapLoad();
} else {
  map.on("load", onMapLoad);
}
```

Tasks

While developers can write code that does a lot with JavaScript, there are some tasks that are best left to a server. The ArcGIS JavaScript API sends requests to the server through **task** objects. There are tasks for calculating the area and perimeter of complex shapes, tasks for querying and returning spatial and non-spatial data on features in a map, and tasks for creating `.pdf` documents containing the map we're looking at, among many others. These tasks take the burden off the browser to perform complex calculations that could be very slow on mobile devices. They also allow the library to be lighter in weight, so the library doesn't have to load every single set of conversion factors between one coordinate system and another, for example.

Most tasks have three parts: the task object, the task parameters, and the task results. The task object lets us send requests to the ArcGIS Server for specific tasks, and includes the constants that may need to be referenced for some tasks. The task object typically takes in a URL string parameter that tells the task to which ArcGIS Server service endpoint to send its requests. The task parameters object defines what information we need to get information from the task. Finally, after we execute the task with the task parameters, and we receive the response from the server, we get a structured object, or a list of objects known as the task's result. Formats for the task, task parameters, and task results can be found in the ArcGIS JavaScript API documentation at `https://developers.arcgis.com/javascript/jsapi/`.

In our code, we're going to use a task called an `IdentifyTask` method. We'll tell the `IdentifyTask` method to contact ArcGIS Server through our census URL. Inside the map click event handler, we'll create a task parameter object called an `IdentifyParameter` object. We'll configure it with the data about the point we clicked, and data on the map. Finally, we'll execute the `IdentifyTask` method, passing in the `IdentifyParameters` object, to retrieve census data from the location we clicked:

```
layer = new ArcGISDynamicMapServiceLayer(censusUrl),
    iTask = new IdentifyTask(censusUrl);

function onMapClick (event) {
  var params = new IdentifyParameters();
  params.geometry = event.mapPoint;
  params.layerOption = IdentifyParameters.LAYER_OPTION_ALL;
  params.mapExtent = map.extent;
  params.returnGeometry = true;
  params.width = map.width;
  params.height= map.height;
  params.spatialReference = map.spatialReference;
  params.tolerance = 3;
  iTask.execute(params);
}
```

Deferreds and promises

Years ago, when FORTRAN ruled the computer programming world, there was one statement that drove developers crazy when it came time to troubleshoot their code: GOTO. With this line of code, disrupting the flow of the application, the application would jump to another line of code. Jumping from one section of code to another made following the application's logic difficult at best.

With modern JavaScript and asynchronous development, following the logic of some asynchronous applications can be difficult, too. The app fires one event when the user clicks a button, which triggers an AJAX request for data. On the successful return of that data, another event fires, which makes the map do something that takes a little time. After the map finishes, it fires off another event, and so on and so forth.

Dojo responded to this by creating **Deferreds** objects. Deferred objects return a promise that a result from an asynchronous process will be coming soon. The function waiting for the result will not be called until that promise is fulfilled. With functions returning Deferred results, the developer can chain functions together using a .then() statement. The .then() statement launches the first function in its parameters only after the result is fulfilled. Multiple .then() statements can be chained together with functions that return Deferred objects, leading to an orderly, and more easily readable coding logic.

In our onMapClick function, the IdentifyTask object's execute method returns a Deferred object. We'll store that deferred result in a variable, so that it can be used by another tool later:

```
function onMapClick (event) {
  var params = new IdentifyParameters(),
      defResults;
  ...
  defResults = iTask.execute(params);
}
```

Showing the results

Once we have received our data, we should show the results to the user. If we look at the format of the data passed over the network, we'll see a list of complicated **JavaScript Object Notation (JSON)** objects. That data, in its raw form, would be useless in the hands of the average user. Thankfully, the ArcGIS JavaScript API provides tools and methods for turning this data into something more user-friendly.

The Map's infoWindow

In the day of modern map applications such as Google Maps, Bing Maps, and OpenStreetmaps, users have been taught that if you click on something important on a map, a little box should pop up and tell you more about that item. The ArcGIS JavaScript API provides a similar popup control for the map called an infoWindow control. The infoWindow highlights feature shapes on the map, and overlays a popup window to show the features-related attributes.

The infoWindow can be accessed as a property of the map (for example, map.infoWindow). From this point, we can hide or show the popup. We can tell the infoWindow which features to highlight. The infoWindow provides a number of configurable and control points to help create a better user experience.

In our application's map click handler, we will need to convert the search results into a form that can be used by the infoWindow. We'll do that by adding an .addCallback() call to the IdentifyTask object's execute function. We'll pull out the features from IdentifyResults, and make a list of them. From that point, we can pass the processed results into the infoWindow object's list of selected features. We'll also prompt the infoWindow to show where the user clicked:

```
function onIdentifyComplete (results) {
  // takes in a list of results and return the feature parameter
  // of each result.
  return arrayUtils.map(results, function (result) {
    return result.feature;
  });
}

function onMapClick (event) {
  ...
  defResults =
  iTask.execute(params).addCallback(onIdentifyComplete);
  map.infoWindow.setFeatures([defResults]);
  map.infoWindow.show(event.mapPoint);
}
```

You can run the application now from your browser, and try clicking on one of the black-outlined features on the map. You should see a shape outlined in cyan (light blue), and a popup pointing to where you clicked. The popup will tell you that there is at least one record there (probably more). You can click the small back and forward arrows on the popup to flip through the selected results (if there is more than one).

The InfoTemplate object

As of this point, we can see the shape of our data. The problem is, we can't see what's inside. There is important tabular data associated with the shapes we're seeing in the results, but the popup hasn't been told how to show the information. For that, we can use an InfoTemplate object. An InfoTemplate object tells the popup how to format the data for display, including what title to use, and how we want the search results displayed. An InfoTemplate object is connected to the feature data, along with the feature's geometry and attributes.

An InfoTemplate object can be constructed in different ways, but most commonly with a string to describe the title, and another string to show the content. The content can contain any valid HTML, including tables, links, and images. Since the title and content are templates, you can insert feature attributes within the template string. Surround the field names from your results with ${fieldname}, where "fieldname" is the name of the field you want to use. If you want to show all the field names and values in the content, without any special formatting, set the content value of the InfoTemplate object to ${*}.

For our application, we'll need to add InfoTemplates to the IdentifyTask results. We'll work with the onIdentifyComplete callback and insert them there. We'll start by inserting the following code:

```
function onIdentifyComplete (results) {
  return arrayUtils.map(results, function (result) {
    var feature = result.feature,
        title = result.layerName;
    feature.infoTemplate = new InfoTemplate(title, "${*}");
    return feature;
  });
}
```

In this bit of code, we're extracting the layer name of the results, and using that for the title. For the content, we're using the "show everything" template to show all fields and values. If you run the web page in your browser now, and click on a feature, you should see something like the following image:

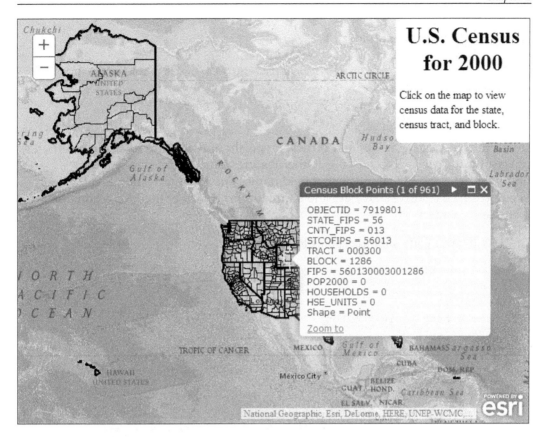

Now, there are a lot of unreadable field names, and field values we may not be able to understand. We could apply different content formats, based on the layer names of the features. Looking at the preceding example, we're mostly interested in the population, the number of households, and the number of housing units:

```
function onIdentifyComplete (results) {
  return arrayUtils.map(results, function (result) {
    var feature = result.feature,
        title = result.layerName,
        content;

    switch(title) {
      case "Census Block Points":
        content = "Population: ${POP2000}<br />Households:
        ${HOUSEHOLDS}<br />Housing Units: ${HSE_UNITS}";
        break;
```

```
      default:
        content = "${*}";
    }
    feature.infoTemplate = new InfoTemplate(title, content);
    return feature;
  });
}
```

If you run the page again in your browser, you should see more readable results, at least for the census block points. Other features, such as the states, counties, and block groups, will show a list of field names and their corresponding values, separated by a colon (:). I'll leave the templates for the other fields as a homework exercise for you.

In the end, your code should read as follows:

```
<!DOCTYPE HTML>
<html>
<head>
  <meta http-equiv="Content-Type" content="text/html;
  charset=utf-8" />
  <meta http-equiv="X-UA-Compatible" content="IE=Edge" />
  <meta name="viewport" content="initial-scale=1,
  maximum-scale=1,user-scalable=no"/>
  <title>Census Map</title>
  <link rel="stylesheet"
  href="http://js.arcgis.com/3.13/esri/css/esri.css">
  <style>
    html, body, #map {
      border: 0;
      margin: 0;
      padding: 0;
      height: 100%;
    }
    .instructions {
      position: absolute;
      top: 0;
      right: 0;
      width: 25%;
      height: auto;
      z-index: 100;
      border-radius: 0 0 0 8px;
      background: white;
      padding: 0 5px;
    }
```

```
    h1 {
      text-align: center;
      margin: 4px 0;
    }
  </style>
  <script type="text/javascript">
    dojoConfig = { parseOnLoad: true, isDebug: true };
  </script>
  <script src="http://js.arcgis.com/3.13/"></script>
</head>
<body>
  <div class="instructions">
    <h1>U.S. Census for 2000</h1>
    <p>Click on the map to view census data for the state,
    census tract, and block.</p>
  </div>
  <div id="map"></div>
  <script type="text/javascript">
    require([
      "esri/map",
      "esri/layers/ArcGISDynamicMapServiceLayer",
      "esri/tasks/IdentifyParameters",
      "esri/tasks/IdentifyTask",
      "esri/InfoTemplate",
      "dojo/_base/array",
      "dojo/domReady!"
    ], function (
      Map, ArcGISDynamicMapServiceLayer,
      IdentifyParameters, IdentifyTask, InfoTemplate,
      arrayUtils
    ) {
      var censusUrl =
      "http://sampleserver6.arcgisonline.com/arcgis/rest/services/
      Census/MapServer/",
          map = new Map("map", {
            basemap: "national-geographic",
            center: [-95, 45],
            zoom: 3
          }),
          layer = new ArcGISDynamicMapServiceLayer(censusUrl),
          iTask = new IdentifyTask(censusUrl);

      function onIdentifyComplete (results) {
        return arrayUtils.map(results, function (result) {
```

```
        var feature = result.feature,
            title = result.layerName,
            content;
        switch(title) {
          case "Census Block Points":
            content = "Population: ${POP2000}<br />Households:
            ${HOUSEHOLDS}<br />Housing Units: ${HSE_UNITS}";
            break;
          default:
            content = "${*}";
        }
        feature.infoTemplate = new InfoTemplate(title, content);
        return feature;
      });
    }

    function onMapClick (event) {
      var params = new IdentifyParameters(),
          defResults;
      params.geometry = event.mapPoint;
      params.layerOption = IdentifyParameters.LAYER_OPTION_ALL;
      params.mapExtent = map.extent;
      params.returnGeometry = true;
      params.width = map.width;
      params.height= map.height;
      params.spatialReference = map.spatialReference;
      params.tolerance = 3;

      defResults =
      iTask.execute(params).addCallback(onIdentifyComplete);
      map.infoWindow.setFeatures([defResults]);
      map.infoWindow.show(event.mapPoint);
    }

    function onMapLoad() {
      map.on("click", onMapClick);
    }

  map.addLayer(layer);

  if (map.loaded) {
    onMapLoad();
  } else {
    map.on("load", onMapLoad);
```

```
        }

    });
    </script>
  </body>
</html>
```

Congratulations, if this is your first web map application using ArcGIS Server and JavaScript. You've worked through multiple steps to produce a working, interactive map application. This application should look great on all the latest browsers. However, older browsers may not connect with the server properly, and may require a proxy.

A note on proxies

If you've worked with the ArcGIS JavaScript API for any length of time, or had to support older browsers like Internet Explorer 9 or less, then you've probably come across proxies. Proxies are server-side applications that make web requests on behalf of the web browser, often transmitting and receiving data the browser could not collect on its own. There are three primary reasons why a browser would require a proxy to communicate with a server. They are as follows:

1. The browser is an older browser that does not support **Cross Origin Resource Sharing** (**CORS**), and the server requests will be made to a different server from the one the application sits on.

2. The proxy provides additional security keys to access specific data, which might include secured tokens the developer doesn't want to release to the public.

3. The web request is much longer than the 2048+ character maximum for GET requests.

The first example is common with older browsers, including Internet Explorer 9 or lower. They can't grab data from third party servers that are separate from the web server, because of a security restraint. With the HTML5 specification for CORS, newer browsers can check to see if an application is allowed to be requested from a script not on the server.

The second example is common in large secure environments, with many security hoops to jump through. Department portal websites could access proxies with tokens unique to the department, providing an extra layer of security.

The third example is common when the user is passing large geometries with many irregular vertices. For instance, if you were to use a drawing tool to draw an irregular shape with the free-handed drawer, its vertices are added as you move around on the map. With so many points, and those points requiring a lot of characters to show their latitude and longitude, it's no wonder a shape request might exceed the browser's maximum character length.

The ArcGIS Server proxy is free to download from GitHub (`https://github.com/Esri/resource-proxy`). They have versions that work with Java, .NET, and PHP. Please use the most recent version, and make sure it's properly configured for all the services you'll be using.

> If both your computer and your ArcGIS Server are behind a network firewall, and the ArcGIS Server has unique public and private IP addresses, your network firewall may block proxy connections to your ArcGIS Server. If you see network traffic failing only on older browsers, such as Internet Explorer 8, and only for internal requests, the firewall might be the issue. Contact your network administrator to work out the issue.

Summary

In this chapter, we have learnt the basics of writing a simple web application using the ArcGIS JavaScript API, and put together a map-based application. We learned how to set up the application in the head of the HTML document. We learned how to create a map on a page and how to add layers so that we can view the data. We learned how to retrieve feature shapes and attributes through a task, and how to show that data to the user.

In the next chapter, we'll look more in depth at the tools available in the ArcGIS JavaScript API.

2
Digging into the API

An **Application Programming Interface** (**API**) defines the operations, formats, and data structures available from a networked service. The developer writes code using the API to tell a server to accomplish a task. An API is built on top of an existing programming language, and uses the syntax and features of the language to make communications with a computer easier to code.

At one point, ESRI had three web-based APIs to communicate with ArcGIS Server: Flash, Silverlight, and JavaScript. With the rise of the mobile web and improvements in browsers, the ArcGIS JavaScript API won as the browser of choice. To make the best use of the ArcGIS Server map features, it's best to learn what's available from the JavaScript API. We're going to take a tour through the API, to become familiar with what it has to offer.

Unlike the other chapters in this book, this chapter provides more of a reference than a programming exercise. Short code snippets have been added to show how to use modules within the API. If you don't absorb every little crumb of information out of this chapter, don't worry. Just come back and reread parts of it for reference.

In this chapter, we'll learn about the following topics:

* How maps are created and configured in the ArcGIS JavaScript API
* How to display geographic data from ArcGIS Server through tiled, dynamic, and graphics layers
* The building blocks of graphics on a map
* How to communicate with ArcGIS Server services through tasks
* How packaged UI controls called **dijits** can save development time

For more information about the ArcGIS JavaScript API, including examples that you can build on your own, check out *Building Web and Mobile ArcGIS Server Applications with JavaScript* by Eric Pimpler or *ArcGIS Web Development* by Rene Rubalcava.

Finding the ArcGIS JavaScript API documentation

You can find the documentation for the ArcGIS API for JavaScript by visiting `https://developers.arcgis.com/javascript/jsapi/`. There, you'll find information on the latest version of the API, a list of changes since the previous version, and related documents to help you create apps. The organization is logical, and information is relatively easy to find with their layout.

The documentation layout

The documentation for each element in the ArcGIS API is laid out in an organized fashion. Links to the API components are on the left-hand side. The documentation for each element defaults to the AMD style, but provides links to the older legacy development style. With the AMD style, the top of the page shows how to load the module into your code. Following that is a description of the module, a link to samples that use the module, and a class hierarchy diagram showing the modules the current module inherits.

Object constructor

If the module requires a JavaScript object to be constructed, the documentation provides information on what is needed to call the module's constructor. The documentation provides details on the parameters required to create the objects, including parameter names, object types, whether they're optional or not, and what they do.

CSS classes and data- attributes

Following any constructor information, the documentation provides data on the CSS class attributes for anything that the module may show on the map, as well as any HTML5 data-* attributes the module might employ. The CSS classes help you to modify the appearance of widgets and visual elements on the module, by giving you a class hook to modify with your custom styling. The data-* attributes give other JavaScript scripts access to information about your widget, without having to load the whole widget into memory. For instance, by clicking on a map graphic while using another library, you could access the geometry's type by looking at the element's data-geometry-type attribute.

Properties

When an ArcGIS API object is created using the constructor, it will have certain properties, or variable parameters assigned to the object. Some properties are set when the object is created. Other properties are set as the object is modified, or certain methods are run. The API documentation lists object properties that are actively maintained, and will not be removed without sufficient notice of depreciation.

If you use web developer tools in your browser, such as Firebug for Firefox, or Chrome Developer tools, you might find that these ArcGIS API objects have more properties than are listed in the documentation. Many properties considered private will have one or two underscore (_) characters before the name. If the properties and methods are not listed in the documentation, they may not be there when the next version of the API is released. Don't count on undocumented properties and methods in your code.

Methods

Besides certain properties, objects created with the ArcGIS API modules will probably contain methods, which are JavaScript functions assigned to the object, often using other parts of the object they're assigned to. The documentation will list public methods attached to the object, as well as what parameters they accept, what values they return, and a description of what task or function they perform.

Events

Many of the modules have documented events, where the module emits a notification that something has happened. These are similar to a HTML button's onClick event, or a browser's onLoad event. In older versions of the API, events were listened to using dojo.connect. But as the library has matured, Dojo is depreciating the connect function and replacing it with the **dojo/on** library (http://dojotoolkit.org/reference-guide/1.10/dojo/on.html). The dojo/on module actually delivers a function.

With the dojo/on function, you attach events by calling it with the following parameters: the HTML DOM elements or Dojo widgets you want to listen to, the name of the event as a string, and a function that will be called when the event occurs. If you want to stop the event, you need a variable to accept the returned value of the dojo/on call, ahead of time. When you're ready to stop listening, call that return variable's remove() method. Some modules contain their own on() method, which implements the dojo/on module for event handling. Here's an example code snippet using the dojo/on module:

```
require(["dojo/on", "dojo/dom", "dojo/domReady!"],
function (on, dom) {
  var soupMe = on(dom.byId("deliver-soup"), "click", function () {
      alert("Here's your soup. NEXT!");
  });
  on(dom.byId("no-more-soup"), "click", function () {
      alert("NO SOUP FOR YOU!");
      soupMe.remove();
  });
});
```

Now that we know what to expect when we're reading the documentation, let's look at the different tools available in the ArcGIS API for JavaScript. Since ArcGIS applications work with geographic data, we'll start off by looking at the map.

Drawing a map

When working with the ArcGIS JavaScript API, one of the first things you'll want to do is to create a map. The map transforms a simple div element into an interactive canvas where you can display, request, and even edit geographic data. The map is a complicated object, so we'll spend a significant amount of time focusing on it. Let's learn how to make a map in our scripts.

Constructor options

When a map is first created using the ArcGIS JavaScript API, the developer has numerous parameters that can be used to configure the map. We were exposed to some of those parameters in the previous chapter, when we created our first map application. Let's look into a few of them a little more closely.

Autoresize

The **Autoresize** function, which by default is true, tells the map whether it should resize itself when the browser changes size or orientation. If you have a full screen map application, and you maximize or resize your browser, the map will automatically adjust and fill the space assigned to it. The map will keep the same zoom level, but more or less only the edges of the map will be visible when you resize your browser.

If your application hides, moves, or animates your maps' appearance onto or off the screen, the autoresize functionality could lead to some strange behavior in the map. For example, if the map is made to slide in from the right, the mouse scroll and some zoom functionality may zoom to the center-left point, rather than the center of the map. With map animations on and off screen, it is best to set the autoresize to false.

Basemap

The **basemap** is an optional string value that provides background imagery to lay your data on. ESRI serves a variety of background imagery, depending on how you want to frame your data. All of ESRI's basemaps are tiled images published in the Web Mercator projection (WKID 102100). For best performances, layers added on top of an ESRI basemap should be in the same projection. Otherwise, ArcGIS Server will have to reproject dynamic map services, and tiled map services may not show up if the spatial references do not match. You can see an example highlighted from the previous chapter in the following code snippet:

```
var map = new Map("map", {
  basemap: "national-geographic",
  center: [-95, 45],
  zoom: 3
});
```

The background imagery options available from ESRI include:

- `street`: An ESRI street map
- `satellite`: Satellite/Aerial photography
- `hybrid`: Combination of street map and satellite
- `topo`: Topographic map with elevation and contours
- `gray`: Light gray/minimalist background
- `oceans`: Oceans and bathymetry background
- `national-geographic`: National Geographic-themed
- `osm`: OpenStreetMap-themed

Explaining the center, extent, zoom, and scale

The **center**, **extent**, **zoom**, and **scale** parameters can be used to override the other map layer settings and set what area is displayed on the map. The *center* can be either a two number array containing the longitude and the latitude of the center point, or an ESRI point. The *extent* describes in map coordinates which edges of the map should be shown: right, left, top or bottom. If a tiled layer is present, the *zoom* instructs which level of display, from furthest out to furthest in, the map should show. The *scale* tells the map how many units of measure relate to one unit in reality (for example, a scale value of 500 means that 1 inch on the map represents 500 inches in real life).

LODs

When a map loads layers, the **levels of details (LODs)** are defined by the first tiled layer loaded in the map. You can override those LODs when creating a map, by passing an array of objects containing a level (number: 0, 1, 2, and so on), a scale (number: for 1:500 scale, you would use 500), and a corresponding resolution (number). This is helpful if you want to limit the user's ability to zoom way outside the view of your map, or if some of your tiled data has a finer resolution than your basemap will show.

Now that we've learned how adjust the parameters of a map object, let's look at its different parts.

Properties

The map object provides a number of properties that can be used in a mapping application. Some are read-only, and tell the status of the map. Others can be manipulated, and the changes can have significant impact on the map. While there are far more properties, we'll take a look at a few of them.

loaded

After you've created a map object, there are some things you can't do to the map until it has at least one layer **loaded**. When a map loads its first layer, it sets its loaded property to true. Then you can manipulate the map, pan around, search for layers, and modify its behavior.

A common programming error is not testing whether the map is loaded before doing something to it, like adding search result graphics from a query. If you have an extremely fast internet connection, and your ArcGIS Server is right next to you, the map might be loaded right after you add a layer. But, for internet users outside your server room, the map may not load that quickly. It's best to check the loaded property before continuing, as in the following example:

```
if (map.loaded) {
  doSomethingwithMap(map);
} else {
  map.on("load", doSomethingwithMap);
}
```

layerIds

The map object keeps a list of references to the layers added to it. Map layer references are stored as an array of strings called `layerIds`. The first layerId in the list refers to the bottom most layer in the stack, while the last id string in the list refers to the top layer visible on the map. This is handy if you need to search for a particular layer in the list, or if you need to do something to all the layers loaded in the map.

spatialReference

The map's `spatialReference` property refers to the long list of mathematical equations used to represent a round earth on a flat map. This attribute is often (but not always) shared by the map and the layers and graphics contained within. We'll go into more depth studying spatial reference in the *Geometry spatial reference* section.

Methods

The map object has a number of useful methods, which are functions tied into the map object. These methods help you add items to the map, move around, zoom in and out, and manipulate map behavior. Let's take a look at some of them.

Adding, removing, and retrieving layers

Layers can be added and removed from the map as needed. If you want to add one map layer at a time, you can call on the map's addLayer() method. To add multiple layers at once, you have to put all those layers in an array and call the map's addLayers() method. Map layers can be removed with the removeLayer() method.

Retrieving layers is as simple as calling the map's getLayer() method. The getLayer() method is passed an id string for the layer. A list of the valid ids can be found in the map's layerIds array, or in the map's GraphicsLayerId property. If no layers are found that match the getLayer() parameter, the method returns a null.

Moving around the map

The map contains several methods for navigating the map. Some affect the location on the map, and others affect how zoomed in the user is. Here is a list of the most common methods used to manipulate your position on the map:

- setZoom(): This sets the map zoom level, moving further in or further out.
- setScale(): Similar to setZoom(), but tries to change the map's scale to match the value passed.
- setLevel(): Similar to setZoom(), but works only in maps with a tiled service. Based on the integer passed into the function, the map will zoom to the corresponding zoom level, if the map has tiled layers.
- setExtent(): When passed a bounding box, this message will attempt to set the map's boundaries to match the bounding box. This affects both zoom level and position.
- centerAndZoom(): Accepts a center point and a zoom level, then centers the map on that point and attempts to zoom to the zoom level. This affects both position and zoom level.
- centerAt(): When this function accepts a point, it attempts to pan the map over so that the point is in the center of the map.

Please note that, for the `setExtent()`, `centerAndZoom()`, and `centerAt()` methods, the points and bounding boxes need to be in the same coordinate system as the map.

Updating settings

There are many map navigation actions that can be enabled or disabled using map methods. You can control what happens when the user clicks, double-clicks, or shift-double-clicks on the map. You can enable or disable panning with the mouse, as well as scrolling with the scroll wheel. You can even disable an enable-all navigation event for the map. There may be times you want to disable navigation. For instance, if your map is navigated by clicking on links, you may not want the user to wander too far on your map.

Remember that these settings cannot be modified until the map is loaded. But if the map hasn't loaded yet, how are we supposed to know when it loads? For an answer, we'll look into a feature common to most JavaScript libraries, events.

Events

Events are important when working with asynchronous programming in JavaScript. The map object has a number of events it fires before, during, and after it makes some important changes, or finishes certain tasks. These events can be listened to through the map's built-in `on()` function. Let's look at some of the more common events you'll be working with:

- `load`: A map event that fires after the map has loaded its first layer.
- `click`: This fires when the user clicks on the map. Among other event items, it returns the point the user clicked as a `mapPoint` property of the event object.
- `extent-change`: This fires when the map either changes location, or changes zoom level. Both of these factor into calculating the map's extent.
- `layer-add`: This fires when any new layer is added to the map.
- `layers-add-result`: This fires after multiple layers are loaded using the `addLayers()` method.
- `mouse-over`: This fires when the mouse is moved over the map.
- `update-start`: This event fires before layers are loaded, either by adding new layers, or by moving around the map and triggering the map to download more map data.
- `update-end`: This event fires after the map loads the data from its map services.

Layers

No map is complete without data to show. Data is added to the map through layers. Layers refer to data sources that include geographic, symbolic, and even tabular data. The ArcGIS API for JavaScript contains a number of modules for interpreting different kinds of common geographic data sources. Some of the layer modules work with data provided by ArcGIS Server, and others can display data from other sources, such as KML, **Web Map Service (WMS)**, and CSV files.

Common ArcGIS Server layers

ArcGIS Server provides map layer data through Map Services. Map Services provide data published from an ArcMap map document (.mxd). The visual map data, legend, and underlying data tables are served through ArcGIS Server, and can be consumed through the browser when the layer data source is loaded into the map. Different layer types have different functionalities. We will review those layers in the following sections.

ArcGISDynamicMapServiceLayer

An ArcGISDynamicMapServiceLayer is the default type of map published through the ArcGIS Server. It is a dynamic map, meaning that its content is updated whenever the map is refreshed. Because it is dynamic, the data can also be reprojected to fit on another map layer that is not in the same spatial reference.

```
require(["esri/map", "esri/layers/ArcGISDynamicMapServiceLayer"],
  function (Map, ArcGISDynamicMapServiceLayer) {
  …
  var map = new Map("mapdiv", {…});
  var layer = new ArcGISDynamicMapServiceLayer
  ("http://serv.er/arcgis/rest/services/MyDynamicLayer/
  mapserver");
  map.addLayer(layer);
});
```

Because the data within an ArcGISDynamicMapServiceLayer is dynamic, ArcGIS Server redraws what the layer should look like every time the map looks at the data from a different location and zoom level. If this layer contains lots of labels, annotation, complicated graphics, or raster data (like aerial imagery, for instance), ArcGIS Server will require more system resources to render the map layer. This will lead to noticeably slower performance. Other services, like tiled map services, are recommended for that kind of data.

ArcGISTiledMapServiceLayer

ArcGIS Server allows map services to be tiled. An `ArcGISTiledMapServiceLayer` has its content already drawn into images at scales defined when the map service is published. These images are stored on a hard drive for quick retrieval. The benefit is that these pre-rendered tiles are served quickly and with little effort on the part of the server. More map services can be run from the same machine with little effect on performance. See an example in the following code snippet:

```
require(["esri/map", "esri/layers/ArcGISTiledMapServiceLayer"],
    function (Map, ArcGISTiledMapServiceLayer) {
    …
    var map = new Map("mapdiv", {…});
    var layer = new ArcGISTiledMapServiceLayer
    ("http://serv.er/arcgis/rest/services/MyTiledLayer/mapserver");
    map.addLayer(layer);
});
```

Tiled map services have some disadvantages. First, the data displayed through tiled map services doesn't change until the tiles are rebuilt. If a park boundary changes in the data, but the tiles aren't rebuilt, the service will still show the old boundary. Second, tiled map services restrict the map to specific zoom scales, with no easy way to view the map in between zoom scales. Third, if another tiled map is stacked on top of the first one, it not only has to be in the same spatial reference, but also the zoom scales have to overlap exactly within overlapping scale ranges. That means that a map built with scales of 1:500 and 1:1000 can be loaded with a map built with scales of 1:1000 and 1:2000, but a tiled layer with scales 1:750, 1:1500, and 1:3000 will not be seen.

You cannot load a dynamic map service layer as an `ArcGISTiledMapServiceLayer`, or else it will throw an error. However, a tiled map service can be loaded as an `ArcGISDynamicMapServiceLayer`. You lose performance when zooming and panning, but it will help you view scales in between the tiled scales.

We've seen ArcGIS layers that provide image data. Let's turn our attention to layers that handle more vector data.

GraphicsLayers and FeatureLayers

Sometimes you don't want to work with representative images of maps. Sometimes you want to work with the shape of your data directly. That's where GraphicsLayers and FeatureLayers come in. These layers render vector graphics directly in your browser. You can click on them, modify them, and even add data to an ArcGIS service through them. Let's look a little closer at them.

GraphicsLayers

GraphicsLayers provide a simple way to add custom vector graphics to your map. You can manage the graphics by calling add() and remove() methods on those graphics. You can also search through the list of graphics on the map by searching through the GraphicsLayer.graphics array.

As a side note, the map object has a built-in GraphicsLayer. You can access that GraphicsLayer through the map.graphics property.

FeatureLayers

FeatureLayers are specialized layers built on a GraphicsLayer that provide much more functionality. They have the custom vector graphics of a GraphicsLayer, plus they have an ArcGIS Server data source to populate the data. You can query for data by either location or attributes, add it to the map using custom graphics, and control popup content and styling. You can even edit data on your server.

Unlike ArcGIS Tiled and dynamic layers, which can mash up multiple layers at once, each FeatureLayer can only work with one map service layer. FeatureLayers can work with individual layers in a map service, or special editable map services called **Feature Services**. Feature Services provide more details about the map layer, including endpoints for querying results. If ArcGIS Server is configured to allow editing on the features, and the map service is published to allow editing, FeatureLayers that consume these services will let you modify both shape and tabular data.

Because FeatureLayers query data from ArcGIS Server, they are limited in the number of features they can return. Typically, ArcGIS Server limits the number of results returned in a query to one or two thousand, depending on the version. Server settings can be altered to show more results, but more FeatureLayer graphics on the map mean more memory is consumed, and the app may be slow and unresponsive. Careful thought must be put into using the FeatureLayer with your data.

One way that the `FeatureLayer` refines the number of features returned is by setting its mode. The `FeatureLayer` has three primary modes of feature selection:

- `MODE_ONDEMAND`: Only features that fit within the map extent, time extent, and/or definition query are downloaded. This is the most common way of interacting with features visible within the map extent.

- `MODE_SELECTION`: Only features that are selected by interacting with the map (by click, touch, or query, for instance) are downloaded and their data is shown on the map.

- `MODE_SNAPSHOT`: All features are initially downloaded into memory. This method is popular for smaller datasets, and requires fewer server calls.

- `MODE_AUTO`: It automatically picks between `MODE_ONDEMAND` and `MODE_SNAPSHOT` and is fairly new to the ArcGIS API for JavaScript, depending on the size and content of the data.

We will look into `FeatureLayers` more closely in *Chapter 5, Editing Map Data* when we discuss editing.

Other layer types

The ArcGIS API for JavaScript works well with data published through ArcGIS Server. But not all geographic data available in the world is published through ArcGIS Server. The API has a number of other layer types that can consume these different data sources. Some of the more common types are as follows:

- `KMLLayer`: Services written with the **Keyhole Markup Language** (**KML**). KML was made famous by Google Earth. The format uses the markup to store and symbolize geographic data. Many services publish KML data, including the **National Oceanic and Atmospheric Administration** (**NOAA**).

- `CSVLayer`: A fairly recent addition, it transforms a comma-delimited text file into points on the map. To show the data on a map, the `CSVLayer` expects a latitude and longitude column to show the coordinates in decimal places. This tool, combined with HTML5's drag-and-drop API, could make it possible for you to map any compliant data you toss into a `.csv` file.

- `WMSLayer` and `WMTSLayer`: Web Map Services (**WMS**) and Web Map Tiled Services (**WMTS**) published using the **Open Geospatial Consortium** (**OGC**) compliant standards.

Please note that if you load the layer using just the URL of the OGC Web Map Service, you'll need to use a proxy (see *Chapter 1, Your First Mapping Application*). This is because the module first requests GetCapabilities on the layer to grab metadata from the map service. This request requires a proxy. However, if you load the WMSLayer with an optional resourceInfo parameter, that automatically describes the service for you, and the proxy will not be necessary.

- StreamLayer: A layer displaying live streaming data from an ArcGIS GeoEvent Processor Extension. StreamLayer takes advantage of the latest in HTML5 WebSockets, which provide real-time updates from the server to the client. You can view Twitter responses to specific events, or real-time locations of emergency response vehicles with tracking devices.

Note that most up-to-date modern browsers support WebSockets technology, but older browsers typically do not. You can visit http://www.websocket.org/echo.html to see if your browser supports WebSockets.

Graphics

A graphic object represents an individual point, line, or polygon feature drawn on the webmap. The graphic object has four main parts: its geometry, symbol, attributes, and infoTemplate. They are used in many parts of the API. If you draw something on the map, you create a graphic. If you query a map service for something, it returns a list of graphics. Some modules even accept lists of graphics as arguments for other functions.

The graphic object can be constructed with up to four optional arguments:

- geometry: It describes the shape of the graphic drawn on the map
- symbol: It describes the graphic's color, thickness, and features that affect the appearance of the graphic
- attribute: A JavaScript object containing name-value pairs of tabular data that correspond with the feature
- infoTemplate: It formats the look of the graphic attributes when highlighted by the map's InfoWindow

We'll look more closely at these graphic features in the following sections.

Introducing geometry objects

Geometry objects are a collection of coordinates that describe where something is in the world. They give shape to the data we're looking for. They can be used for displaying graphics on a map, or to provide spatial data for queries, or even for geoprocessing services. Geometry objects come in five basic types: point, line, polygon, extent, and multipoint. Any shapes are built on these basic types. Let's take a closer look.

Point

In the ArcGIS API for JavaScript, a point is the simplest geometry object. It contains the x and y coordinates of the point, as well as the point's spatial reference. A point can be constructed with x and y coordinates, plus a spatial reference. It can also be defined by longitude and latitude.

Polyline

A polyline geometry is a collection of one or more arrays of point coordinates. Each array of point coordinates is called a path, and would look like a line when drawn out. Multiple paths can be stored in the same geometry, giving the appearance of multiple lines. The individual point coordinates are stored as array of x and y values.

Polygon

Polygons are made up of one or more arrays of points that loop back on themselves. Each array of points is referred to as a ring, and the first and last point in a ring must have the same coordinate. Polygons made up of more than one ring are referred to as multipart-polygons.

Among the many properties and methods that control the polygon's shape are two useful methods: `.getCentroid()` and `.contains()`, `.getCentroid()` returns a point that is roughly in the middle of the polygon. The formula used to calculate the centroid's position can be found on the Wikipedia page for centroids: (`http://en.wikipedia.org/wiki/Centroid#Centroid_of_polygon`). The `.contains()` method takes a point as an input, and returns a `boolean` value based on whether the point is inside the polygon or not.

Extent

An extent is a rectangular polygon, which only uses four numbers to describe its shape. The extent is described by its minimum *x* value (xmin), maximum *x* value (xmax), minimum *y* value (ymin), and maximum *y* value (ymax). When viewed on a map, extents typically look like a box.

Extents are used by many different parts of the API. All other geometry objects other than points have a getExtent() method, which is used to collect the bounding box of the geometry. You can zoom to a specific location on a map by setting the map's extent. Extents are also used to define an area of interest when identifying things on a map.

Multipoint

Sometimes you need a cluster of points to show on the map. For that, the multipoint is your answer. You can add and remove points using the addPoint() and removePoint() methods. You can also get the general area the points cover with the getExtent() method. They're useful for collecting random points in an area, and for selecting unconnected features on a map.

Geometry spatial reference

Besides coordinates of a geometry's location, the geometry object also packs information on the object's spatial reference. The spatial reference describes the mathematical models used to describe the Earth within the mapped region. Instead of containing all those formulas and constants, the spatial reference stores a reference to those formulas through either a **well-known ID (WKID)**, or a **well-known text (WKT)**.

Spatial Reference plays a crucial role when displaying data on a map. Tiled map service layers must have the same spatial reference as the map, in order to be displayed. Dynamic map service layers must be reprojected, adding to the server time and workload. Graphics also need to be in the same spatial reference as the map to appear in the correct places on the map.

Symbols and renderers

The ArcGIS API provides some basic graphics tools to define how features look on your map. If you are creating your own custom graphics, or modifying some existing ones, this portion of the API tells you how to define their **symbols**. The API also provides custom **renderers**, which help you define what symbols are used, based on rules applied by the renderer.

Simple symbols

The ArcGIS API uses symbols to define the colors, thicknesses, and other visual features of graphics, independent of their shape. In fact, without a symbol assigned, the graphics would not show up. The API defines three simple symbols to highlight points, line, and polygons. Let's look at the Simple Symbols.

SimpleLineSymbol

The `SimpleLineSymbol` defines the symbology of line and polyline graphics added to the map. It may seem strange to start with line symbology, but the other two Simple symbols make use of a line symbol as well. The primary `SimpleLineSymbol`'s properties are its color, its width (in pixels), and its style. Styles include lines that are solid, dashed, dotted, and combinations of those as well.

SimpleMarkerSymbol

The `SimpleMarkerSymbol` defines the symbology of point and multipoint graphics on the map. The symbol can generate different shapes, depending on the style that is set. While this symbol defaults to a circle, the `SimpleMarkerSymbol` can generate squares, diamonds, x's, and crosses. Other properties that can be manipulated within this symbol include the size in pixels, its outline as a `SimpleLineSymbol`, and its color.

SimpleFillSymbol

`SimpeFillSymbols` define the symbology of polygon graphics on the map. The style of the `SimpleFillSymbol` reflects how the inside of the shape is shown. It can be solid (completely filled in), null (completely empty), or have lines running horizontally, vertically, diagonally, or crossing. Please note that a graphic with a null-styled symbol does not respond to clicks in the center. The symbol can also be modified by a `SimpleLineSymbol` outline, and if the style of the `SimpleFillSymbol` is solid, its fill color can be modified as well.

Let's look a little further at the colors we'll use to define these symbols.

esri/Color

The `esri/Color` module allows you to customize the colors of the graphics you add to your map. The `esri/Color` module is an extension of the `dojo/_base/Color` module, and adds some unique tools to blend, create, and calculate color formats in different ways. Older versions of the library used the `dojo/_base/Color` module to set colors, and is still usable in the current version.

The constructor for the esri/Color module can accept a variety of values. You can assign common color names, such as "blue" or "red". You can use hex strings (for example #FF00FF for purple), like you would find in css colors. You can also create colors using three or four number arrays. The first three numbers in the array assign the red, green, and blue values from 0 to 255 ([0, 0, 0] is black, [255, 255, 255] is white). The fourth optional number in the array indicates the alpha, or the opacity of the color. Values in this range are from 0.0 to 1.0, with 1.0 being completely opaque, and 0.0 being completely transparent.

The esri/Color module has some interesting enhancements on the old dojo/_base/Color module. For instance, you can create a blend of two existing colors with the blendColors() method. It takes two colors, and a decimal number between 0.0 and 1.0, where 0.5 is an even blend of the two colors. This could be useful, for instance, on a voting map where green represents yes, and red represents no, and voting districts could be colored by percentages of yes or no votes.

The esri/Color modules also have ways to translate colors from one format to another. For instance, you could click on a feature, get its graphic color, and use toHex() or toCss() to get the color string. From there, you could apply that color to the background color of an information <div> which is used to show the attributes of that graphic.

Picture symbols

If pictures speak to you more than colors, you can use Picture Symbols in your graphics. With these symbols, you can use picture icons and custom graphics to add interesting and memorable data points on your graph. Points and polygons can be symbolized using Picture Symbols, using the modules esri/symbols/PictureMarkerSymbol and esri/symbols/PictureFillSymbol respectively.

The PictureMarkerSymbol module adds simple picture graphics for points on the map. It is typically constructed with three arguments, a URL pointing to the image file (such as .jpg, .png, or .gif), and an integer width and height in pixels.

The PictureFillSymbol module fills the content of a polygon with a repeating graphic. The PictureFillSymbol takes four arguments. The first is a URL string pointing to an image file. The second input is a line symbol, like the esri/symbols/SimpleLineSymbol. The final two arguments are integer width and the height of the image in pixels.

Renderers

Sometimes, you don't want to assign graphic symbols one at a time. If you know what type of graphics you're adding, and how the graphics should look ahead of time, it makes sense to assign the symbology ahead of time. That's where renderers come in. Renderers can be assigned to GraphicsLayers as a way to assign a common style to all graphics inside them. Renderers depend on a graphics layer accepting one geometry type (like all polygons, for instance).

As the ArcGIS API has matured, it has added a number of different renderers to its library. We'll take a look at three common ones: the SimpleRenderer, the UniqueValueRenderer, and the ClassBreakRenderer. Each has its appropriate use cases.

SimpleRenderer

The SimpleRenderer is the simplest of the renderers because it only assigns one symbol. It accepts a default symbol, and assumes that you'll only insert graphics of the same type as the symbol. Like other renderers, if you assign a SimpleRenderer to a GraphicsLayer, you can add Graphic objects to the GraphicsLayer without them having a symbol assigned. Let's look at a code snippet that shows how to create and apply a SimpleRenderer to the map's graphics layer:

```
require(["esri/map", "esri/renderers/SimpleRenderer",
  "esri/symbols/SimpleLineSymbol", "esri/Color",
  "dojo/domReady!"],
function (Map, SimpleRenderer, LineSymbol, Color) {
  ...
  Var map = new Map("map", {basemap: "OSM"});
  var symbol = new LineSymbol().setColor(new Color("#55aadd"));
  var renderer = new SimpleRenderer(symbol);
  ...
  map.graphics.setRenderer(renderer);
});
```

The SimpleRenderer has been expanded since version 3.7 of the API. Symbols can be modified in different ways, based on a range of expected values in a field. For instance, you can vary the symbol's color between two colors by setting the colorInfo property of the SimpleRenderer. You can also change the size of a point according to a field value by setting its proportionSymbolInfo property. Finally, you can rotate a symbol by its field value by setting its rotationInfo property.

Unique Value Renderer

Sometimes you want to display different symbols for your graphics based on one of their attributes. You can do that with a Unique Value Renderer. Specific unique values can be highlighted, while values not in the list can be assigned a default symbol.

The constructor of the Unique Value Renderer accepts a default symbol, and up to three different fields to compare values against. If more than one field is used, a field delimiter string is required. Unique values and the associated symbols can then be loaded as objects within the renderer's info parameter. The info object requires the following:

- A unique value (`value`)
- A symbol associated with that value (`symbol`)
- A symbol label (`label`)
- A description of the unique value (`description`)

Class Break Renderer

Sometimes you want to classify your graphics by where they fit in certain ranges of values, like population or **Gross Domestic Product** (**GDP**). The Class Break Renderer helps you to organize the graphics in a visual manner. This renderer looks at a particular field value in a graphic, and finds which class break range graphic's field value fits between. It then assigns the corresponding symbol to that graphic.

When constructing a Class Break Renderer, you assign a default symbol, and the field that will be evaluated to modify the symbol. Class breaks are assigned through an info property of the renderer. The info array accepts a JavaScript object containing the following:

- The minimum value of the class break (`minValue`)
- The maximum value of the class break (`maxValue`)
- The symbol (`symbol`)
- A label describing the break (`label`)
- A description of the class break (`description`)

InfoTemplates

InfoTemplates, as we mentioned in *Chapter 1, Your First Mapping Application,* describe how you want to display data in a popup. The InfoTemplate can be constructed using two strings, the title and the content. The title and content strings represent HTML templates, and you can substitute graphic attributes in the template. For a graphic's infoTemplate, if you wanted to display the graphic's total attribute, you would insert the value in the template with a ${total}, which shows the field name inserted inside the brackets of the substitution string ${}.

If you want to display a popup containing all the name/value pairs in the graphic's attributes, you can use a wildcard ${*} in the content value, as follows:

```
graphic.infoTemplate = new InfoTemplate("Attributes", "${*}");
```

For the content, you can add HTML content to the string, including tables, lists, images, and links. For example, you might use something like the following to describe a marine habitat:

```
graphic.infoTemplate = new InfoTemplate("Marine Habitat
   (${name})", "<table><tbody><tr><th>Type:
   </th><td>${type}</td></tr><tr><th>Ocean
   zone:</th><td>${zone}</td></tr><tr><td colspan='2'><img
   src='${imageSrc}' alt='${name}' /></td></tr></tbody></table>");
```

In the preceding example, we added the name attribute of the graphic to the title, and created a two-column table that displays the type of the graphic, the zone where the graphic resides, and a two-column wide image connected to the map graphic.

Toolbars

The ArcGIS API for JavaScript uses toolbars to handle controls that transform the cursor. The API has three different toolbars: one for map navigation, one for drawing graphics, and one for editing those graphics. These tools are turned on using the activate() method, which includes parameters describing what tool you want to use. The tools can be turned off using the deactivate() method; let's take a closer look at these tools.

The navigation toolbar

The navigation toolbar handles advanced map navigation. By default, the map offers zoom sliders, and the ability to pan around the map. By activating the zoom in/zoom out features of the navigation toolbar, the user can draw a boxed extent where the map will either zoom in on, or zoom out at the same ratio as the zoom box to the map's current extent. This zoom in/out box is always on once you activate it, so you'll have to either develop something that turns it off, or let the user activate the pan tool, which disables the zoom in/out.

The navigation controls have three more actions that can be activated. The zoomToFullExtent() function, when called from the navigation toolbar, triggers the map to zoom to its original map extent. The toolbar also keeps track of how many zoom-ins, zoom-outs, and pans the user has done. By activating the zoomToPreviousExtent() function, the map extent goes back to previous extents. And if the map goes back to previous extents, the map can also zoom to future locations. With the zoomToNextExtent() function, the user can undo the zoom extent changes they made when they viewed the previous extent.

The draw toolbar

The draw toolbar allows the user to draw on the map to provide input. Depending on which draw tool was activated, the user can draw a shape on the map, and any event listener attached to the draw-end event is called, running a function that receives the shape the user drew. Multiple draw-end events can be attached to the event listener if needed. Common draw-end events might include adding a drawing to the map, measuring the shape drawn by the toolbar, or using the shape drawn in a query. The tools also modify the tooltip of the map, giving directions about how to draw with them.

The ArcGIS API provides numerous shapes to use when drawing on the map. The toolbar can be activated using simple drawing tool constants like point, polyline, and polygon, which are activated by clicking each point on the map and double-clicking to complete. There are freehand versions of the polyline and polygon, which allow you to click and drag the mouse to draw a continuous line or shape, and release the mouse to stop drawing. Also included in the API are drawing tools for common shapes such as triangles, rectangles, circles, ovals, and arrows, just to name a few.

The different toolbars are not aware of each other. If you are not careful, you could activate both drawing and navigation tools at the same time, and their actions will occur simultaneously. You could draw a shape for a tool, and be zoomed into that area because the navigation tool wasn't told to deactivate. Plan ahead how you will disable one event when another is enabled.

The edit toolbar

What happens when you draw something on the map and you have made a mistake? How do you correct a graphic on the map? The edit toolbar allows you to change map graphics' shape, size, and orientation. It also works on text symbols, letting you change the text of a text graphic on the map.

When the edit toolbar is activated on a feature, it allows you to move, rotate, scale, or change the vertices of the feature. Moving lets you grab the feature and position it where you want. Rotating provides a point to click and drag around to re-orient the graphic. Scaling lets the user make the graphic larger or smaller. For editing vertices, the user can drag points around on the map, right-click to "delete" the point, and click between the graphic versions on "ghost points" to add new vertices to the map.

Please note that while the Edit Toolbar allows users to change graphics on the map, it doesn't have any features built in that let the user save their changes. When using the edit toolbar on graphics in an editable feature layer, you still will need a way to tell the feature layer to save changes. Graphics can be edited in the current session without saving the changes.

Tasks

As I mentioned in the previous chapter, some application items require resource-intense processes that are best left for the server. The ArcGIS API calls these actions tasks. There are a large variety of tasks. Some require the server to compute and/or return data, which is the case for the geometry service and the QueryTask. Others use the server to generate custom documents, like the Print Task. Still, others also do all this, and also require custom extensions not initially a part of ArcGIS Server. Let's take a look at what makes a good task.

Tasks, parameters, and results

As we discussed in the previous chapter, a task is made up of three parts, the task constructor, the task parameters, and the task results. Typically, only the task constructor and the task parameters need to be loaded using AMD. The task results are automatically produced when the results return from the server. The format of the results is shown in the API documentation to show you how to access what you need when you successfully receive results. The common order of events with tasks is as follows:

1. The task and task parameters are constructed.
2. The task parameters are filled in with the required information
3. The task executes a server call using the task parameters.
4. When the browser receives the results from the server, something is done with the results either through a `callback` function or a deferred value.

It is worth noting that when tasks are executed, they return one of Dojo's **Deferred** objects. These deferred results can be passed to another function to be run asynchronously when the browser receives the results.

Common tasks

There are many tasks commonly available on any installation of ArcGIS Server. Many rely on a published map service or geoprocessing service to act on. Some, like the `GeometryService`, are built into ArcGIS Server. Let's take a look at some of the common tasks available through the API.

GeometryService

`GeometryService` provides multiple functions for manipulating geometry data. Typically, it is published as part of ArcGIS Server. Let's take a look at the Geometry Service functions available.

- `AreasAndLengths`: Finds areas and perimeters of polygon geometries
- `AutoComplete`: Helps create polygons adjacent to other polygons by filling in gaps between them
- `Buffer`: Creates a polygon whose edges are a set distance or distances from the source geometry
- `ConvexHull`: Creates the smallest polygon shape necessary to contain all the input geometries

- `Cut`: Splits a polyline or a polygon along a secondary polyline
- `Difference`: Takes a list of geometries, and a secondary geometry (probably a polygon), and returns the first list any overlap against the secondary geometry is cut out
- `Distance`: Finds the distance between geometries
- `Generalize`: Draws a similar looking shape with much fewer points
- `Intersect`: Given a list of geometries and another geometry, returns the geometries defined by where the first list intersected with the second
- `Label Points`: Finds a point within a polygon which would be the best position to place a label
- `Lengths`: For lines and polylines, finds the planar or geodesic distance from beginning to end
- `Offset`: Given a geometry and a distance, will create a geometry the defined distance to the right of the original geometry if the distance is positive, or to the left of the geometry if negative
- `Project`: Based on a geometry and a new spatial reference, it returns the original geometry in the new spatial reference
- `Relation`: Based on two sets of geometries and a specified relationship parameter, returns a list of pairings showing how items in the two lists are related (for example: how many from list2 are inside list1)
- `Reshape`: Given a line or a polygon, and a secondary line, it changes the shape of the first geometry based on the second one, adding to the original figure, or tossing some of it away
- `Simplify`: Given a complicated polygon or line whose edges crosses itself, it returns simpler polygons where the crossed parts are undone
- `TrimExtend`: Given a list of polylines, and a polyline to trim or extend to, it cuts lines in the list where they cross over the second polyline, and extend other polylines where they don't quite reach the secondary polyline
- `Union`: Given a list of polygons, creates a unified geometry where all the polygons are dissolved along edges that overlap, creating one solid polygon

QueryTask

The QueryTask provides a way to collect both shape and attribute data from features in a map service. You can request data on individual layers in a map service, defining your search area by geometry and/or limiting your results to attribute queries. The QueryTask is constructed with a URL parameter pointing to a single layer in a map service. The number of results returned from a successful query is limited by ArcGIS Server, typically to either 1,000 or 2,000 results.

Because results for a QueryTask can be limited by ArcGIS Server, the QueryTask object has been extended with multiple methods to collect data on its search results. Here are some of the methods available through the QueryTask:

- execute(): Accepts the QueryTask parameters, and requests features based on those parameters.

- executeForCount(): Accepts the same parameters, and returns the number of results in the map service layer that match the query.

- executeForExtent(): It is based on the parameters passed, returns the geographic extent that contains all the search results. Note that this is only valid on hosted feature services on www.arcgis.com.

- executeForIds(): With the parameters passed, the results return a list of ObjectIds for the features that match the search. Unlike the execute command, where the number of results are limited by ArcGIS Server, the executeForIds() method returns all the ObjectIds that match the search. That means, if there are 30,000 results that match the search, this method will return 30,000 ObjectIds, instead of the server limit of 1,000.

The QueryTask parameters can be created using the Query module (esri/tasks/query). If you have experience with SQL, you'll see similarities between the Query parameters and SQL Select statements, while there are some Query parameters that are more specific to ArcGIS Server. A few of the common ones are as follows:

- where (string): A legal where clause for a query.

- geometry (esri/geometry): A shape to spatial select against.

- returnGeometry (Boolean): Whether you want the geometry of the search result features. Setting this to false returns just attribute data.

- outFields (array of strings): A list of fields to retrieve the search results. Assigning ["*"] to the outFields parameter returns all fields, which is equivalent to the SQL statement SELECT * from.

- `outSpatialReference` (`esri/SpatialReference`): The spatial reference you want the geometry to be drawn in. Typically, the map's spatial reference is assigned to this value.

- `orderByFields` (array of strings): A list of field values to sort the results by. Add `DESC` to the string if you want values returned in descending order.

Results are returned from `QueryTask` in a form called `FeatureSet` (`esri/tasks/FeatureSet`). `FeatureSet` contains data about the results returned, including the geometry type of the results, the spatial reference, field aliases for the results, and a list of feature graphics that match the search results. The `featureSet` even returns whether or not the number of results that match the query exceeded the number that could be delivered by ArcGIS Server. The array of features within the features property of the `featureSet` can symbolized and added straight to a map's graphics layer to be seen and clicked.

IdentifyTask

The `IdentifyTask` provides a shortcut way to retrieve data from a map service for a popup. Unlike the `QueryTask`, which queries only one layer in the map service, the `IdentifyTask` can search through all layers in the map service. While the `QueryTask` has numerous methods to retrieve data on a search, the `IdentifyTask` is more simplified, with a single `execute()` command to retrieve server results.

The parameters passed to an `IdentifyTask` are referred to as `IdentifyParameters`. Because `IdentifyTasks` can search multiple layers, the `IdentifyParameters` focus on searching by geometry. To search by a `where` clause, a `layerDefinition` parameter must also be added to the parameters.

The results of an `IdentifyTask`, known as the `IdentifyResults`, differ somewhat from the `FeatureSet`. First, `IdentifyResults` are returned as an array of objects that contain a feature parameter that contains the result graphic, while the `FeatureSet` is a single object with an array of graphics in its features parameter. Secondly, `IdentifyResults` do not contain a list of field names and aliases like the `FeatureSet`, because the feature attributes of the `IdentifyResult` are name/value pairs where the name key is the alias. Finally, the `IdentifyResults` return the layer id and name of the layer they were retrieved from. These attributes makes the `IdentifyTask` favourable to quickly populate a map's popup on a click.

FindTask

The ArcGIS API offers another tool for finding features on a map called a `FindTask`. The `FindTask` accepts a text string, and searches for features in the map service that contain that value. On the database side, the `FindTask` queries through each layer in the map service, searching permitted fields for a text string with wildcards before and after it. If searching for the string "Doe", the `FindTask` would return a feature with the name value of "John *Doe*", "*Doe* Trail", or "`Fiddledee` *doe* Studios", but would not return a feature containing "Do exit" because of the space.

Locator

A Locator provides either an approximate location for an address based on street addressing data, known as **geocoding**. The locator can also perform **reverse geocoding**, which provides an approximate address for a point on the map based on that same street data. ESRI provides geocoding services on a national and world level, but if you have more up-to-date street information, you can publish your own geocoding service on ArcGIS Server.

For address location, the locator accepts a parameter object containing either a single line input, or an object containing a Street, City, State, and Zone (zip code) parameter. The `callback` function that accepts the `Locator.execute()` results should accept a list of objects called `AddressCandidates`. These `AddressCandidates` include the possibly matching street address, the point location of the address, and a score from 0 to 100, where 100 is a perfect match. If no results return there were no matches for the address provided.

Tasks requiring server extensions

While ArcGIS Server packs a lot of functionality, ESRI also provides server extensions, or optional software that performs specific tasks through ArcGIS Server. This software can do a number of unique tasks, from providing driving directions to sifting through tweets in your area. You will need to know if your server has one or more of these extensions before trying to use them. We'll take a look at one of them, the Routing Task.

Routing

Routing Tasks can provide driving directions, calculate the most effective routes, and redirect users around known roadblocks. The Routing Tasks require the Network Analyst extension for ArcGIS Server. For Routing Parameters, you can add a `FeatureSet` into the stop parameter, and fill it with graphics defining the various stops our vehicle has to make. You can also supply barriers in the same way, which block certain roadways. When the Routing Task is executed, it returns a best route based on the street data provided to the service.

Dijits

The ArcGIS API for JavaScript also provides Dojo widgets, commonly referred to as **dijits**. Many of these dijits offer a **user interface** (**UI**) with existing API modules working in the background. Using these dijits can cut development time by providing well-used and tested UI components. Dijit documentation also shows the CSS classes used to style the dijits, so that developers and designers can restyle them to fit the theme of the page. Let's take a look at the more commonly used dijits.

Measurement

The Measurement dijit provides a tool that can be used to measure distances, areas, and perimeters of locations on the map. It can also get the latitude and longitude of points on the map. The dijit uses the drawing toolbar to draw shapes on the map. From there, it sends a request to a `GeometryService`. Once the dijit has retrieved its results, it either displays the latitude and longitude of a point, shows the length of the line drawn on the map, or displays the area and perimeter of the polygon area drawn on the map.

Print

The Print dijit provides a dropdown control that lets the user select from a predefined list of print templates and parameters. The developer configures the dijit with a link to an ArcGIS Server print service, and a list of predefined print parameters. When the user selects from one of the dropdown choices, the `printTask` incorporated in the Print dijit fires off an `execute()` method with the corresponding print parameters. When the `printTask` receives a success response from the server, the dijit provides a link to click on to download the printout.

If you need more granularity in your print control, this dijit is not for you. This tool is not meant to handle every combination of print parameters. If you're required to support everything from Tabloid ANSI B Landscape `.pdf` documents to Letter ANSI A Portrait `.gif` images, and everything else in between, you should consider developing your own tool.

Bookmarks

The Bookmarks dijit allows the user to save custom areas on the map. You can pre-assign bookmarked areas in the configuration file that the user can click and zoom to. They can also add areas to a list, edit the names, and delete the custom areas they want to zoom to. The Bookmarks dijit does not save the map state, meaning that the layers that were turned on previously won't be automatically switched, and the selected graphics on the map may not be there.

Basemaps

The ArcGIS API provides a couple dijits for changing your basemaps. This can be useful when the data blends in or doesn't look right against the current background. The first tool the API provides is a `BasemapToggle`, which lets you switch between two basemaps. The second is the `BasemapGallery`, which offers more choices. Both provide thumbnail photos and captions describing the basemaps.

The `BasemapGallery` can be customized to show your custom basemaps, or show basemaps from ArcGIS Online. When constructing the `BasemapGallery`, the choice of incorporating ArcGIS Online basemaps can be modified by setting the `showArcGISBasemaps` option to true. Remember the rule about tiled map services, however. The ArcGIS Online basemaps are in Web Mercator Auxiliary Sphere (WKID 102100), which is the same projection as Google and Bing Maps. If your data is not in the same projection, you could have issues with accuracy and/or missing tiled layers.

Popups and InfoWindows

Popups and `InfoWindows` provide a small window to view data about features on a map. There are four varieties of the `infoWindows` available: The legacy version that was replaced in version 3.4, a smaller version of the legacy `infoWindow` called `InfoWindowLite`, the current default popup control, and a lighter mobile version of the current popup.

The current popup exposes a number of elements that you can tap into to customize the user experience. The popup can select multiple features, and holds the graphics content of the selected items in a features array. It also has a paging control that lets the user click to see what other features have been selected. It also keeps track of which feature is selected through a `selectedIndex` parameter.

Editing

The editing dijits provide UI components and tools that you can use to edit features on the map. These tools work with editable `FeatureLayers`, allowing the user to edit shapes, add features on the map, change attributes, and save all those changes to the server. Let's look at some of the components that make up the editing tools.

AttributeInspector

The `AttributeInspector` allows the user to edit graphic attribute data from an HTML form. The inspector connects to the `FeatureLayer` and displays data on one of the features that is selected. The inspector also provides a previous and next button to search through the selected Features. The `AttributeInspector` also supplies a delete button to delete selected features.

The inspector then builds a form based on the requirements of the features. Dates will be given calendars, and domain-bound fields will show an HTML Select element with a dropdown list of domain choices. The user can type in and edit data, and the data can be saved server-side.

TemplatePicker

The `TemplatePicker` provides a selector to select what feature types and subtypes you'll be editing. When a `featureLayer` is published as editable, it has to expose the symbology and preset styles through the REST service. The `TemplatePicker` then reads the exposed symbology and feature setup data to create a grid of valid editable choices. The user clicks on the templates like a button, and the symbology is passed to a drawing tool of some kind.

An all-in-one Editor

The ArcGIS API provides an all-in-one **Editor** dijit to provide some basic editing needs. The dijit incorporates the `TemplatePicker` to select features types and subtypes to draw from. It also has a toolbar with various tools, pertaining to the editable features on the map. So, when you select a green line, the tool will show line-related drawing tools.

The Editor also generates an `AttributeInspector` in the `map.infoWindow`. You can edit field values from the popup. It is up to you how you implement saving changes.

The editor also provides undo and redo buttons in the toolbar. If you accidently delete a feature you wanted to save, you can use the undo button. If you realize that you really wanted to delete that feature, you can click the redo button.

Now that you have been exposed to most of the major features of the ArcGIS API for JavaScript, you are better prepared to write some more code using the API.

Summary

In this chapter, we've taken a whirlwind tour of the ArcGIS API for JavaScript. We've gained familiarity with how the website and API is organized. We've studied the Map object in detail, looking at all its parts. We've looked over the other parts of the API, including layers, graphics, toolbars, tasks, and dijits. We've discussed how each of these are implemented, and some use cases for them. Now that we know more about the parts of the API, we can use these components to create custom applications and widgets in the next chapter.

3

The Dojo Widget System

The developers at Esri created the ArcGIS JavaScript API using the Dojo framework. Dojo provides a large assortment of tools, libraries, and UI controls that work across multiple browsers. Any developer can create custom applications with UI elements that work well together using Dojo and the ArcGIS JavaScript API. Also, Dojo provides the AMD tools necessary to develop your own custom widgets, libraries, and controls.

In the previous chapters, we've reviewed the ArcGIS API for JavaScript, and we've written a small application using the API. We've even incorporated some basic principles of AMD into a single page application. What we've done so far would work great with a smaller application.

But what happens when applications get larger? Are we going to implement a single script to load all the components we need for a larger application? How are we going to extend the functionality of our website? What if we update our libraries and something breaks? Are we going to hunt through several thousand lines of code in one monolithic JavaScript file to find the parts that need to be changed?

In this chapter, we'll take advantage of the Dojo framework and create a custom widget for our application. Through that process, we'll cover the following:

- The background of the Dojo framework
- What the `dojo`, `dijit`, and `dojox` module packages have to offer
- How to create and use our own custom modules
- How to create widgets (modules with UI components), and how to extend them

A brief history of the Dojo framework

The Dojo framework began in 2004 with initial work at Informatica. Alex Russell, David Schontzler, and Dylan Schiemann contributed the first lines of code for the project. As work on the code continued, other developers were brought in and lent their input into the direction of the framework. The project grew so large that the founders created the Dojo Foundation to oversee the codebase and its intellectual properties. Since then, over 60 developers have contributed to the framework, and companies such as IBM and SitePen continue to use it today. For more information, visit at `http://dojotoolkit.org/reference-guide/1.10/quickstart/ introduction/history.html`.

So what makes Dojo a framework, as opposed to a library? When this question was posed to the people at stack overflow (`http://stackoverflow.com/ questions/3057526/framework-vs-toolkit-vs-library`), the most agreed upon answer centered on an inversion of control. When we use tools in a library, our code controls the flow of logic and activity. When we calculate a value, we update that value in each location that it's needed through our code. In a framework, however, the framework controls the behavior of the application. When a value is updated in a framework, it is the framework that updates that value wherever it is bound on the web page.

The Dojo framework provides a number of HTML-based controls that we can load and use. Much in the way of CSS appearances and JavaScript behavior, is prewired into the control. After initial setup, our code instructs the framework where to send data when specific control events occur. We have no control when our functions are called, only what happens after they're called.

 If you want to read more about frameworks and the inversion of control, Martin Fowler provides a good explanation of the subject on his blog at `http://martinfowler.com/bliki/InversionOfControl.html`.

Introducing dojo, dijit, and dojox

Great care has been taken to organize the Dojo framework. When Dojo incorporated the AMD style of modules, many of the objects, methods, and properties were reorganized into logical folders and modules. Dojo is broken down into three main packages: **dojo**, **dijit**, and **dojox**. Let's get an idea of what these three packages bring to the framework.

The dojo package

The dojo package provides much of the base functionality needed to load, run, and tear down the various modules in our applications. The modules provide functions and tools that work across multiple browsers, including the dreaded older versions of Internet Explorer. For example, the developer doesn't have to handle events by trying addEventListener() for Chrome and attachEvent() for older IE, because it's handled behind the scenes by dojo/on. With the framework, you can get away from all the browser hacks, and focus on creating a good application.

The dojo package does not contain widgets, but it does contain the tools necessary to manipulate things within the widgets. Do you need to handle mouse clicks and touch events? Use the dojo/mouse and dojo/touch modules. Do you need a jQuery-like selector for your HTML? Load the dojo/query module. The dojo package provides modules for HTML DOM manipulation, arrays, AJAX, dates, event, and i18n, just to name a few.

The dijit package

The dijit package provides visual elements (referred to as **dijits**) that integrate well with the dojo package. The elements created with the dijit package have been tested across multiple browsers. They provide a consistent user interface based on the CSS stylesheets loaded with the library.

As the Dojo framework is used and contributed to by many companies and developers, there are plenty of user controls available. Whether you're creating a submission form, a calendar of events, or an executive dashboard, you can combine the dijits in this package in a way that suits your needs. Some of the more popular packages include:

- dijit/Calendar: This provides an interactive HTML calendar control that works on both desktops and tablets.
- dijit/form: A package containing stylized form elements such as buttons, checkboxes, dropdown menus, textboxes, and sliders. The form elements have a consistent look across older and newer browsers.
- dijit/layout: A package containing controls that handle layouts. You can create simple containers, tab containers, accordion containers, and even containers that control the positions of other containers.
- dijit/Tree: A module that creates a collapsible tree menu which could be used, for instance, to show folder layouts.

The dijit package contains more than user controls. It also provides the tools necessary to create your own custom dijits. Using the dijit/_WidgetBase module and assorted mixins, you can incorporate both HTML elements and existing dijits into your own custom dijit. Working with the dijit components can give the user a consistent experience throughout the entire application.

The dojox package

According to documentation on the Dojo framework (http://dojotoolkit.org/reference-guide/1.10/dojox/info.html), dojox provides extensions to Dojo. This part of the library deals with more experimental functionality, user interfaces, and testing features. Many parts of the library have matured due to extensive use, while other parts have been depreciated and are not under active development.

One of the useful subpackages in the dojox package is dojox/mobile. The Dojox/mobile package provides UI elements and controls that you can use in mobile applications. They have been tested on a wide variety of mobile browsers, and their styling can even mimic the style of native smartphone and tablet applications.

> For more information regarding the dojo, dijit, and dojox packages in the Dojo framework, you can view tutorial documentation by going to: http://dojotoolkit.org/documentation/tutorials/1.10/beyond_dojo/.

The dojoConfig packages

Using the built-in Dojo packages is great and all, but what about making your own custom package? You can create custom packages in your local folders, and reference them through your dojoConfig object. In the dojoConfig object, you can add a packages parameter that contains an array of JavaScript objects. Those package objects should contain a name attribute, which is a reference to your package, and a location attribute, which references the folder location. Here's an example of a dojoConfig object with a reference to a custom package:

```
<script>
  dojoConfig = {
    async: true,
    isDebug: true,
    packages: [
      {
        name: "myapp",
```

```
            location: location.pathname.replace(/\/[^/]+$/, '') +
            "/myapp"
        }
    ]
  };
</script>
```

In the preceding sample, the package `myapp` is referenced, and all the files for that package are loaded into the `myapp` folder under the current page. So, if this page was shown at `http://www.example.com/testmap/`, the `myapp` package could be found at `http://www.example.com/testmap/myapp`. When referencing the `Something` module in your `myapp` package, you would load the module like this:

```
require(["myapp/Something", "dojo/domReady!"], function
    (Something) { … });
```

Defining your widget

With Dojo's AMD style, there are two main ways to use AMD components. Using the `require()` function plays the script once, and then it's done. But if you want to create a module that can be used over and over again, you would want to `define()` the module, instead. The `define()` function creates one or more custom modules to be used by an application.

The `define()` function takes a single argument, which could be any JavaScript object. Even `define("Hello World!")` is a valid module definition, though not that useful. You can create more useful ones by passing in objects or object constructors that perform tasks for your application. Review the following example:

```
define(function () {
  var mysteryNumber = Math.floor((Math.random() * 10) + 1);
  return {
    guess: function (num) {
      if (num === mysteryNumber) {
        alert("You guessed the mystery number!");
      } else if (num < mysteryNumber) {
        alert("Guess higher.");
      } else {
        alert("Guess lower.");
      }
    }
  };
});
```

In the preceding example, the module picks a random number from one to ten, It then returns a module with a `guess()` method that accepts a numeric value. Upon calling the `guess()` method, it alerts the user as to whether the guess was correct or not.

Declaring modules

Developers entering JavaScript from more classical **Object-oriented (OO)** languages may have a difficult time embracing the language's prototype-based inheritance. In most OO languages, classes are defined at compilation, and sometimes inherit properties and methods from other base classes. In JavaScript objects, class-type objects are stored within the object's prototype, which holds the same shared properties and methods for multiple related objects.

Dojo can ease the transition for these developers when they use the `dojo/_base/declare` module. The module creates Dojo-based classes, which can be used by other applications. Behind the scenes, it takes a class-like JavaScript object and converts it to a prototype-based constructor. This module couples well with the `define()` function to create custom `dojo` modules, like the following:

```
define(["dojo/_base/declare", …], function (declare, …) {
  return declare(null, {…});
});
```

The `declare()` function takes three arguments: the class name, the parent class, and the class properties object. The class name is an optional string that can be used to reference the class. The `declare()` function will turn the name into a global variable, so that the `dojo/parser` can write it into HTML when you reference the `dijit` package with a `data-dojo-type` attribute. If you do not plan to use the `dojo/parser` to write your widget into HTML, it's highly recommended that you don't use the first argument.

Class Inheritance through Dojo

The second argument in the `declare()` function refers to the parent class. There are three possible ways to define the parent class. If the class you are creating does not have a parent class, it is said to have no inheritance. In that case, the parent class argument is null, as in the following statement:

```
define(["dojo/_base/declare", …], function (declare, …) {
  return declare(null, {…});
});
```

In the second scenario, there is a parent class for the class you are creating. The parent could be another AMD module, or another declared item. The following example shows this by extending the `dijit/form/Button` module:

```
define(["dojo/_base/declare", "dijit/form/Button", …],
  function (declare, Button, …) {
    return declare(Button, {…});
  }
);
```

The third possible scenario involves multiple inheritance, where your class inherits properties and methods from multiple classes. For that, the parent class parameter will also accept an array of objects. The first item in the array is considered the base class, which provides the main construction parameters. The rest of the items in the list are referred to as "mixins". They don't necessarily provide object construction functionality, but they add properties and methods that either complement, or override the existing base class. The following example shows multiple inheritance using a few libraries we'll talk about later:

```
define(["dojo/_base/declare", "dijit/_WidgetBase", "dijit/_
OnDijitClickMixin, "dijit/_TemplatedMixin"],
  function (declare, _WidgetBase, _OnDijitClickMixin,
  _TemplatedMixin) {
    return declare([_WidgetBase, _OnDijitClickMixin,
  _TemplatedMixin], {…});
});
```

Dealing with classes and inheritance

In the final argument of the `declare()` statement, the developer defines the different properties and methods contained in the class. The object properties can be anything from strings and numbers to lists, objects, and functions. Please review the following code example:

```
declare("myApp/Fibonacci", null, {
  list: [],
  contructor: function () {
    var len = this.list.length;
    if (len < 2) {
      this.myNumber = 1;
    } else {
      this.myNumber = this.list[len - 1] + this.list[len-2];
    }
    this.list.push(this.myNumber);
  },
  showNumber: function () { alert(this.myNumber); }
});
```

In objects created with the `declare()` statement, you have both static and instance properties. **Static** properties are properties that are shared by all objects that were created in the same way. **Instance** properties are properties that are unique to the object created. Properties defined within the class object are considered static, but every property initially assigned either within the constructor, or within another method, is considered an instance property.

In the preceding example, the `showNumber()` method and the `list` properties are static, while the `myNumber` property is an instance. That means every `myapp/Fibonacci` object will share the `showNumber()` method and the `list` array, but will have unique `myNumber` properties. For example, if five `myapp/Fibonacci` objects are created, each should contain the list value of `[1, 1, 2, 3, 5]`. Adding one more to the list will add to every Fibonacci list. The `myNumber` property is created in the constructor, therefore each object will have a unique value for that property.

When a class is created from a parent class, it has access to its parent's properties and methods. The new class can also overwrite those methods by having a new method with the same name in its constructor. For example, let's say a `HouseCat` object inherits from a `Feline` object, and they have their own versions of the `pounce()` method, If you call `HouseCat.pounce()`, it will only run the method described in `HouseCat`, and not in `Feline`. If you want to run the `Feline.pounce()` method within the call to `HouseCat.pounce()`, you would add `this.inherited(arguments)` within the `HouseCat.pounce()` method, to show when you want to run the parent class method.

Working with Evented modules

The `Evented` module lets your widgets publish events. When your module is declared with `Evented` as the parent class, it provides an `emit()` method for broadcasting that the event has occurred, and an `on()` method for listening to events. One example that can be found in the ArcGIS API would be the drawing toolbar. It doesn't display information, but it has the tools necessary to publish events.

The `emit()` method takes two arguments. The first is a string name that describes the event, such as `map-loaded` or `records-received`. The second is an object created and passed along with the event. The object could be anything you can create in JavaScript, but remember to keep the returned content similar, so that the methods listening for the event to occur don't miss it.

An overview of the _WidgetBase module

The _WidgetBase module provides the base class necessary to create a dijit module. In itself, it's not a widget, but you can easily build a widget on top of it. The widget created with the _WidgetBase module is bound to an element in the HTML DOM, which can be referred to with its domNode attribute.

The _WidgetBase module also introduces a lifecycle to the widget. The lifecycle refers to how the widget is created, built up, used, and torn down. This module loads the following methods and events that occur within the lifecycle of the widget:

- constructor: This method is called on widget creation. There is no access to the template, but values can be assigned and arrays can be accessed.

- parameters mixed into widget instance: The parameters you passed into the object constructor, such as button labels, are added to the widget to overwrite any pre-existing values.

- postMixinProperties: This is a step prior to rendering the widget HTML. If you need to change or correct any values passed into the parameters or something, this would be the time to do so.

- buildRendering: This is where the HTML templates are added in place of the existing node, if templating is being used. If you're using the _TemplatedMixin module, the template string is loaded into the browser, rendered into HTML, and inserted in place of the existing domNode. Events bound in the HTML data-* attributes are assigned here, as well.

- Custom setters are called: If you have added custom setter functions, these will be called at this point.

- postCreate: Now that the HTML has been rendered, this function can do more with it. Just be warned, the widget HTML may not be connected to the document yet. Also, if this widget contains child widgets, they may not have rendered yet.

- startup: This function can be called after all parsing and child-widget rendering has finished. It is usually used when the widget needs to resize itself through the resize() method.

- destroy: This function is called when this widget is torn down and removed, either by closing the application, or whenever this function is called. The parent classes typically handle the tear down events on their own, but if you need to do anything unique before destroying the widget, this would be the function to extend.

The `_WidgetBase` module has specialized getter and setter functions that allow you to perform specific tasks when you set a widget property. You can retrieve the values with the `get()` function and set these properties using the `set()` function. The first argument of both `get()` and `set()` is the name of the property, and the second argument in the `set()` method is the value. So, if you want to set the `name` of a widget, you would call `widget.set("name", "New Name")`. There is also a `watch()` method that can perform functions when a value changes through `set()`.

Working with other _Mixins

The `_WidgetBase` module may not provide all the widget functionality your application will need. The `dijit` package provides mixins, or JavaScript object extensions, for your widget. These mixins can provide HTML templates, event handling for clicks, touches, focus, and a lack thereof, for example. Loading the right mixin with your application can save you a lot of behavior coding. Let's take a look at some of the ones we may use.

Adding _TemplatedMixin

The `_TemplatedMixin` module lets the module replace its existing HTML content with either a string template, or HTML from another source. The widget's template string is assigned through the `templateString` property. This allows the widget to skip a possibly complicated `buildRendering` step. You can see an example of a `_TemplatedMixin` module call in the following code snippet:

```
require(["dojo/_base/declare", "dijit/_WidgetBase",
  "dijit/_TemplatedMixin"],
function (declare, _WidgetBase, _TemplatedMixin) {
  declare("ShoutingWidget, [_WidgetBase, _TemplatedMixin], {
    templateString: "<strong>I AM NOT SHOUTING!</strong>"
  });
});
```

As the `templateString` property is being rendered, properties can pass back and forth between template and widget. Widget properties can be written to the template during the `buildRendering` phase by referencing the property name in the widget, and surrounding it with the `${}` wrapper. You can also assign node references and attach widget events using the HTML data attributes. The `data-dojo-attach-point` attribute lets the widget have a named property that connects to the template. The `data-dojo-attach-event` attribute attaches a `callback` method in the widget that is triggered when the event is occurs. You can view the following example:

```
<div class="${baseClass}">
  <span data-dojo-attach-point="messageNode"></span>
  <button data-dojo-attach-event="onclick:doSomething">
    ${label}
  </button>
</div>
```

 How should a developer lay out a template widget? The widget's template HTML should all be contained in a single HTML element. You can use a `<div>` tags, a `<table>` tags, or element tags you plan to enclose the template in. If the template contains more than one element at the base layer, the widget will not render and throw an error.

Adding _OnDijitClickMixin

The `_OnDijitClickMixin` module provides the ability for elements within your template to be "clicked". This works, not only with click events using the mouse, but through touch and keyboard events as well. Besides clicking and touching an element, the user can tab until the element is highlighted, then press the *Enter* or spacebar key to activate it.

If the `_OnDijitClickMixin` module is loaded, the developer can add event handlers to the `dijit` template through the `data-dojo-attach-event` attribute. Within this data attribute text value, add `ondijitclick:` followed by the name of your click event handler. You must have this event handler pointing to a valid function, or else the whole widget will fail to load. An example of a `dijit` template using the `clickMe(event) {}` function is as follows:

```
<div class="clickable" data-dojo-attach-
  event="ondijitclick:clickMe"> I like to be clicked.</div>
```

As a side note, the function within the click event handler should be ready to accept a one click event argument, and nothing else.

Adding _FocusMixin

The `_FocusMixin` module provides focus and blur events for your widget and its components. For instance, if you have a widget that takes up more room when it's focused, you could add an `_onFocus()` event handler to the object like this:

```
_onFocus: function () { domClass.add(this.domNode, "i-am-huge"); }
```

On the other hand, when you want your widget to shrink back to its normal size, you would add an _onBlur() event handler:

```
_onBlur: function() { domClass.remove(this.domNode, "i-am-huge");}
```

The event system

The event system within JavaScript is an important part of the language. JavaScript was designed to respond to events in the browser and events caused by users. A website may need to respond to the click of a user, or an AJAX response from a server. With the language, you can attach functions called event handlers to listen for events from certain elements in the page and the browser.

In the Dojo framework, events are listened to using the dojo/on module. When assigning an event listener, the module function call returns an object that lets you discontinue listening by calling the object's remove() method. Also, dojo/on has a once() method that fires the event once, when the event occurs, then automatically removes the event.

Some older ArcGIS API samples still use Dojo's old event handler, dojo.connect(). Events handlers would be attached with dojo.connect(), and removed with dojo.disconnect(). The Dojo Foundation is currently depreciating dojo.connect(), and will drop it from their code when Dojo moves to version 2.0. If you are maintaining old code, please start migrating all dojo.connect() to dojo/on. With the ArcGIS API, please pay attention to event names and results returned. Names may change from CamelCase to dash-separated, and while dojo.connect() can return more than one item in its callback, dojo/on only returns a single JavaScript object.

Events are created using emit(). This function takes in a string name of the event, and a JavaScript object you would want to pass to an event listener. The emit() method is available for widgets using the dojo/Evented module, and those built using the dijit/_WidgetBase module. It is also available through the dojo/on module, but dojo/on.emit() requires an HTML DOM before the event name.

Now that we have a grasp of the Dojo framework, let's build something with it.

Creating our own widget

Now that we know the basics of creating custom widgets, that is to say, dijits for our web applications, let's put our knowledge into practice. In this part of the chapter, we will transform the single-page application code we wrote in *Chapter 1, Your First Mapping Application* into a dijit that can be used with any map. We'll then use that `dijit` in a map application that can be expanded to include other dijits as well.

Picking up where we left off...

We finally received feedback from the Y2K Society about the census map we created in *Chapter 1, Your First Mapping Application*. They liked how the map worked, but felt it could use more polish. After a meeting with some group members, here is a list of their feedback:

- The application should have sections at the top and the bottom to show titles and a little bit about the society
- The colors should be warm and cheery, such as yellows
- Get rid of the plain black line work of the census layer data
- The application should let us do more than simply look at census data. It should let us click on other buttons to do other things
- The census information needs to be more organized and grouped logically in tables

In order to comply with their requests, we will need to reorganize our application from *Chapter 1, Your First Mapping Application*. To fix the issue, we'll need to do the following:

1. We will create a new full page app with a header and footer.
2. We will modify the style of the application.
3. We will represent the census map service using custom symbology.
4. We will add buttons to the header to launch the census popups.
5. We will move the `infoTemplate` data into separate HTML files.

The file structure for our application

In order to deliver a map that can do more than just show census data, we're going to use the tools provided by the Dojo framework. We'll use elements of Dojo to lay out the application, and turn our previous census map tool into its own widget. Instead of a single page application, where the HTML, CSS, and JavaScript are mashed together in the same file, we're going to separate the files into distinct components.

Let's start by organizing the project folder. It's a good idea to separate the styling and scripts into separate folders. In our project folder, we'll add two subfolders to the project folder. We'll name them css and js, for the stylesheet and script files, respectively. Inside the js folder, we'll add another folder and name it templates. This is where our widget templates will go.

In the main project folder, we'll create an HTML page called index.html. Next, In the css folder, we'll create a file called y2k.css, which will be our application stylesheet. We'll create two JavaScript files in our js folder: one called app.js for our application, and another called Census.js for our widget. We'll stub out template files in the js/templates folder for the census widget (Census.html), plus some templates for our popups. The files should look something like the following folder structure:

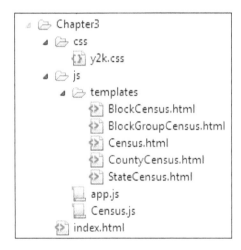

Defining the layout with Dojo

To handle the layout of the application, we'll use a couple of the modules within the dijit/layout package. This package has a number of UI containers with styled looks and built-in UI behavior that we don't have to reinvent. We'll add the layout elements and restyle them to fit our client's wishes. Let's look at the layout module found within the dijit package.

Framing the page with BorderContainer

The BorderContainer container provides a common layout tool used with Dojo. The BorderContainer container is extended from the LayoutContainer module, which creates a layout that fills up the entire browser window without scrolling. The LayoutContainer module (and by extension, the BorderContainer module) allows content elements to be placed with a region attribute. The BorderContainer container adds borders, spacing, and resizing elements called splitters, which can be dragged to resize the content items.

The regions that can be assigned to content dijits inside the BorderContainer container are as follows:

- **Center**: This positions the element at the center of the page, relative to the other items. One element must be assigned this value.
- **Top**: This positions the element above the center element.
- **Bottom**: This positions the element below the center element.
- **Right**: This positions the element to the right of the center element.
- **Left**: This positions the element to the left of the center element.
- **Leading**: If the dir attribute of the page is set to "ltr" (left-to-right), the leading element is placed to the left of the center element. If the dir attribute is "rtl" (right-to-left), the leading element is placed on the right.
- **Trailing**: Layout is the opposite of the leading element, with "ltr" on the right, and "rtl" on the left.

The BorderContainer module, and its parent LayoutContainer module, have a design parameter that affects the containers surrounding the center element. When the design parameter is set to its default value of headline, the top and bottom panels take up the entire width of the container. The right, left, leading, and trailing panels shrink to fit between the top and bottom panels. Alternatively, you can set the design parameter to sidebar, and the panels on the sides will take up the entire height, while the top and bottom panels are squeezed in between them. The following figure shows example pages made with headline and sidebar designs. The design parameter was the only parameter changed in the examples:

BorderContainer with 'headline' (left) and 'sidebar' (right) design attributes with ContentPanes

Inserting ContentPane

The ContentPane tile provides the general container used in most desktop Dojo pages. ContentPane tiles can adjust to fit within their assigned places, and can expand and shrink as the user resizes the browser. ContentPane tiles can house HTML elements, such as other Dojo widgets and controls. They also have an href attribute which, when set, will load and render the content of another HTML page. The pages are loaded via **XMLHttpRequest (XHR)**, therefore the loaded HTML should be on the same domain, due to cross-origin issues.

Modifying the layout of our application

Looking at the requirements, we're going to use the BorderContainer containers and some ContentPane tiles to create the header, footer, and the content of the application. We'll take a standard HTML template pulling the necessary libraries, then add the dijit content.

In your application's folder, start by creating an HTML file called index.html. We'll title it Y2K Map, and add the necessary meta tags and scripts to load the ArcGIS API for JavaScript. We'll also load the dojo CSS stylesheet nihilo.css, to handle the styling of the base styling of the dojo elements in the application. In order to apply the style to our application, we'll also add a class of nihilo to the body element of our HTML document. Finally, we'll add a link to our y2k.css file, which we'll create soon. Your HTML document should look like the following:

```
<!DOCTYPE HTML>
<html>
  <head>
    <meta http-equiv="Content-Type" content="text/html;
    charset=utf-8" />
    <meta http-equiv="X-UA-Compatible" content="IE=Edge" />
    <meta name="viewport" content="initial-scale=1,
    maximum-scale=1,user-scalable=no" />
    <title>Y2K Map</title>
    <link rel="stylesheet"
    href="http://js.arcgis.com/3.13/dijit/themes/nihilo/
    nihilo.css">
    <link rel="stylesheet"
    href="https://js.arcgis.com/3.13/esri/css/esri.css" />
    <link rel="stylesheet" href="css/y2k.css" />
    <script src="https://js.arcgis.com/3.13/"></script>
  </head>
  <body class="nihilo">
  </body>
</html>
```

Next, we'll add the BorderContainer container and the ContentPane tiles inside the body tag. These will be built on basic <div> elements. We'll give the BorderContainer container an id of mainwindow, a ContentPane tile at the top with an id of header, another ContentPane tile with an id of map where our map will go, and a ContentPane tile at the bottom with an id of footer. We'll also add a little content to the header and footer, just to make it look nice. Here is an example:

```
<body class="nihilo">
  <div id="mainwindow" data-dojo-type="dijit/layout/BorderContainer"
  data-dojo-props="design:'headline',gutter:false,liveSplitters:
  true" style="width: 100%; height: 100%; margin: 0;">
    <div id="header" data-dojo-type="dijit/layout/ContentPane"
      data-dojo-props="region:'top',splitter:true">
      <h1>Year 2000 Map</h1>
```

```
        </div>
        <div id="map" data-dojo-type="dijit/layout/ContentPane"
          data-dojo-props="region:'center',splitter:true">
        </div>
        <div id="footer" data-dojo-type="dijit/layout/ContentPane"
          data-dojo-props="region:'bottom',splitter:true"
            style="height:21px;">
          <span>Courtesy of the Y2K Society</span>
        </div>
      </div>
    </body>
```

Since the client wanted a place to add multiple functions, including our census search, we'll add a location for buttons in the upperright-handcorner. We'll create a button containing <div>, and insert our census button as a dijit/form/Button module. Using the dijit button will ensure that the styling of the part will go along with the styling of the widget. Here is an example:

```
<div id="header"
  data-dojo-type="dijit/layout/ContentPane"
  data-dojo-props="region:'top',splitter:true">
    <h1>Year 2000 Map</h1>
    <div id="buttonbar">
      <button data-dojo-type="dijit/form/Button"
        id="census-btn" >Census</button>
    </div>
</div>
```

To make the files work, we'll need to add a link to the main script file we'll run for our app. We'll insert the app.js file as the last element inside the <body> tag, so that loading the file doesn't cause the browser to block other downloads. In the following code, you can see where we insert this:

```
        <span>Courtesy of the Y2K Society</span>
      </div>
    </div>
    <script type="text/javascript" src="js/app.js"></script>
  </body>
```

Within the `app.js` file, we're going to do just enough to get the visuals. We'll add to this file as we go. We'll start with our normal `require()` statement, loading the modules for the `BorderContainer`, `ContentPane`, and `Button` elements. We'll also pull in a module called `dojo/parser`, which will parse the Dojo data-markup in the HTML and turn it into application widgets. The code will be as follows:

```
require([
  "dojo/parser",
  "dijit/layout/ContentPane",
  "dijit/layout/BorderContainer",
  "dijit/form/Button",
  "dojo/domReady!"
], function(
  parser
) {
  parser.parse();
});
```

After all that work, our HTML should look something like the following:

```
<!DOCTYPE HTML>
<html>
  <head>
    <meta http-equiv="Content-Type" content="text/html;
    charset=utf-8" />
    <meta http-equiv="X-UA-Compatible" content="IE=Edge" />
    <meta name="viewport" content="initial-scale=1,
    maximum-scale=1,user-scalable=no" />
    <title>Y2K Map</title>
    <link rel="stylesheet"
    href="http://js.arcgis.com/3.13/dijit/themes/nihilo/
    nihilo.css">
    <link rel="stylesheet"
    href="https://js.arcgis.com/3.13/esri/css/esri.css" />
    <link rel="stylesheet" href="css/y2k.css" />
    <script src="https://js.arcgis.com/3.13/"></script>
  </head>
  <body class="nihilo">
    <div id="mainwindow"
    data-dojo-type="dijit/layout/BorderContainer"
      data-dojo-props="design:'headline',gutter:false,
      liveSplitters: true"
      style="width: 100%; height: 100%; margin: 0;">
        <div id="header"
```

```
        data-dojo-type="dijit/layout/ContentPane"
        data-dojo-props="region:'top',splitter:true">
        <h1>Year 2000 Map</h1>
        <div id="buttonbar">
          <button id="census-btn"
          data-dojo-type="dijit/form/Button">Census</button>
        </div>
      </div>
      <div id="map" data-dojo-type="dijit/layout/ContentPane"
      data-dojo-props="region:'center',splitter:true">
        <div id="census-widget"></div>
      </div>
      <div id="footer"
        data-dojo-type="dijit/layout/ContentPane"
        data-dojo-props="region:'bottom',splitter:true"
          style="height:21px;">
      <span>Courtesy of the Y2K Society</span>
    </div>
  </div>
  <script type="text/javascript" src="js/app.js"></script>
</body>
</html>
```

Styling our application

If we run the application at this point, it won't look quite as expected. In fact, it'll look a lot like a white screen. That is because we haven't assigned a size to our page yet. In this case, the ContentPane tile styling creates positioned-looking panels, and absolute positioning takes the content out of the calculated page flow. Since there was nothing else to fill the height of the body, it collapsed to zero height.

A quick remedy to this issue is to update the styling of the HTML and the body tag. Open up y2k.css in your text editor and add the following lines of CSS:

```
html, body {
  border: 0;
  margin: 0;
  padding: 0;
  height: 100%;
  width: 100%;
  font-size: 14px;
}
```

This CSS we applied makes the page fill up 100 percent of our browser window. Adding the border, margin, and padding takes away any possible formatting that different browsers may insert into the page. We added the font size because it's good practice to set the font size in pixels at the body level. Further font size assignments can be made relative to this, using the em unit.

If you play your page now, you should see the following output:

Year 2000 Map

Census

Courtesy of the Y2K Society

It's definitely progress, but we need to fulfil the other requirements of the application. First, the #buttonbar element would look better if we positioned it precisely in the top-right, and left a little gap for the **Census** button to be centered vertically. Next, we'll add some yellow hues to the different panels, and round off the corners of the header and footer. We'll end up with the following stylized shell for our application:

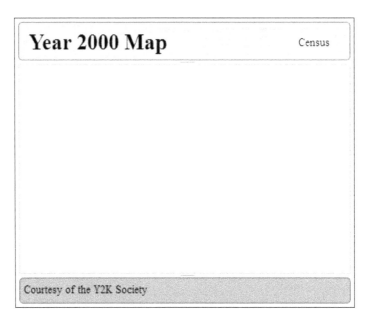

To make this happen, here's the CSS we'll add to our y2k.css:

```css
h1 {
  margin: 2px 8px;
  display: inline-block;
  *display: inline; /* IE 7 compatible */
  zoom: 1; /* IE 7 compatible */
}

#buttonbar {
  position: absolute;
  top: 10px;
  right: 15px;
  width: auto;
  height: auto;
}
```

```css
#mainwindow,
#mainwindow .dijitSplitter {
  background: #fffaa9; /* paler yellow */
}

#header, #footer {
  -moz-border-radius: 5px;
  -webkit-border-radius: 5px;
  border-radius: 5px;
  border: 1px solid #6f6222;
}

#header { background: #ffec50; /* bold yellow */ }

#footer { background: #d0a921; /* darker yellow */ }
```

Adding our custom package

Before we can use our custom package, we need to tell Dojo where it is. To do that, we'll need to add a `dojoConfig` object before we load the ArcGIS JavaScript API. In the `dojoConfig` object, we'll add a package array object, and tell it that the `y2k` modules are in the `js` folder. The script should look something like this:

```html
<link rel="stylesheet" href="css/y2k.css" />
<script type="text/javascript">
  dojoConfig = {
    packages: [
      {
        name: 'y2k',
        location: location.pathname.replace(/\/[^\/]*$/, '') +
          '/js'
      }
    ]
  };
</script>
<script src="https://js.arcgis.com/3.13/"></script>
```

Setting up our app

Now we need to set up our application script and get things moving. We'll open our app.js file and add the functionality for our map and our upcoming Census widget. We'll start by adding references to the esri/map module and our y2k/Census widget. We'll initialize our map just as we did in *Chapter 1, Your First Mapping Application,* and we'll create a new Census widget. Following the pattern of many dojo widgets, the Census widget constructor will take an options object for the first parameter, and a string reference to an HTML node within the <div> map as the second argument. We'll fill in the options later:

```
require([
  "dojo/parser",
  "esri/map", "y2k/Census",
  "dijit/layout/ContentPane",
  ...
], function(
  parser, Map, Census
) {
  parser.parse();

  var map = new Map("map", {
    basemap: "national-geographic",
    center: [-95, 45],
    zoom: 3
  });

  var census = new Census({ }, "census-widget");

});
```

 Perhaps you are wondering why some of the modules in our code start with capital letters, while others do not. Common JavaScript coding convention states that object constructors start with capital letters. The Map and Census modules make maps and census widgets, therefore they should be capitalized. Why the reference to the esri/map module isn't capitalized is a mystery, and a source of errors if you get it wrong.

Coding our widget

We need to start piecing together our Census widget. To do that, we'll use what we've learned about creating custom widgets from earlier in the chapter. Within the Census.js file, we'll create a shell widget using the define(), declare(), and _WidgetBase modules. It should look something like the following code:

```
define([
  "dojo/_base/declare",
  "dijit/_WidgetBase"
], function ( declare, _WidgetBase ) {
  return declare([_WidgetBase], { });
});
```

For our widget, we'll want a template that instructs the user how to use the tool. It might also be a good idea to let the user close the tool, since we don't want the map to be cluttered with multiple tools:

```
define([
  "dojo/_base/declare",
  "dijit/_WidgetBase", "dijit/_TemplatedMixin",
  "dijit/_OnDijitClickMixin"],
  function (declare, _WidgetBase, _TemplatedMixin,
  _OnDijitClickMixin) {
  return declare([_WidgetBase, _TemplatedMixin,
  _OnDijitClickMixin], { });
});
```

At this point, we need to load our template into our widget. To do that, we're going to implement another dojo module called dojo/text.

Adding some dojo/text!

The dojo/text module lets the module download any sort of text file as a string. The contents can be HTML, text, CSV, or any related text-based file. When loading the file using AMD, the format is as follows:

```
require([…, "dojo/text!path/filename.extension", …],
function (…, textString, …) {…});
```

In the preceding example, the filename.extension describes the file name, such as report.txt. The path shows the location of the file in relationship to the script. So a path of ./templates/file.txt means that the file is located in the templates folder, which is a subfolder of the folder where this widget script has been located.

The exclamation point in our declaration means that the module has a plugin attribute that can automatically be called on load time. Otherwise, we would have to wait and call it after the module loaded in our script. Another module where we see this is dojo/domReady. The exclamation point activates that module, pausing our application until the HTML DOM is ready as well.

Getting back to our application, it's time to load our dijit template. The _TemplatedMixin module provides a property called templateString, which it reads from to build the HTML portion of the dijit package. We'll use the dojo/text module to load HTML from our Census.html template, then insert the string created from that into the templateString property. It should look something like this:

```
define([
  ...
  "dijit/_OnDijitClickMixin",
  "dojo/text!./templates/Census.html"
], function (..., _OnDijitClickMixin, dijitTemplate) {
  return declare([...], {
    templateString: dijitTemplate
  });
});
```

Our dijit template

For our template, we're going to take advantage of a few cool tricks in Dojo. The first trick is that we can mix property values into our widget through the same substitute templating we learned about in *Chapter 1, Your First Mapping Application*. Second, we're going to take advantage of the _OnDijitClickMixin module to handle click events.

For our template, we're looking at creating something with a title (such as "Census"), some instructions, and a close button. We'll assign a close button event handler through the data-dojo-attach-event attribute using ondijitclick. We'll also assign a baseClass attribute from the widget to the CSS classes in the widget. If you haven't created the Census.html file in your template folder, do so now. Then, enter the following HTML content:

```
<div class="${baseClass}" style="display: none;">
  <span class="${baseClass}-close"
    data-dojo-attach-event="ondijitclick:hide">X</span>
  <b>Census Data</b><br />
  <p>
    Click on a location in the United States to view the census
    data for that region.
  </p>
</div>
```

What errors in the template could possibly cause the widget not to load? If you are incorporating events, through the `data-dojo-attach-event` parameter, make sure the callback function in the template matches the name of the callback function in your dijit. Otherwise, the dijit will fail to load.

Back in our code, we'll assign a `baseClass` property, as well as functions for `hide()` and `show()`. Many dijits use those functions to control their visibility. Through those functions, we'll set the display style attribute to either `none` or `block`, as shown in the following code:

```
define([…, "dojo/dom-style"],
function (…, domStyle) { …
{
  baseClass: "y2k-census",

  show: function () {
    domStyle.set(this.domNode, "display", "block");
  },

  hide: function () {
    domStyle.set(this.domNode, "display", "none");
  }
});
```

Working with the dijit constructors

We'll need to add onto the `dijit` constructor function, but first we need to think about what this dijit needs. Our original purpose for the dijit was to let the user click on the map and identify census locations. So, we'll need a map, and since we're identifying something, we'll need to supply a map service URL to the constructor of the `IdentifyTask` object.

Following the trends of most `dijit` constructors, our `dijit` constructor will accept an options object, and either an HTML DOM node, or a string `id` for that node. In the options object, we'll look for a map, and a `mapService` URL. The second argument in the constructor will be assigned as the dijit's `domNode`, or if it's a string, the appropriate node will be found based off the string.

For our constructor, we'll incorporate the map into the widget, and turn the `mapService` into an `IdentifyTask`. We'll also add the `dojo/dom` module to provide a shortcut for the common JavaScript operation `document.getElementById()`, and use that to transform a string `id` into a DOM element:

```
define([…, "dojo/dom", "esri/tasks/IdentifyTask", …],
function (…, dom, IdentifyTask, …) {
  …
  constructor: function (options, srcRefNode) {

      if (typeof srcRefNode === "string") {
        srcRefNode = dom.byId(srcRefNode)
      }

      This.identifyTask = new IdentifyTask(options.mapService);
      this.map = options.map || null;
      this.domNode = srcRefNode;
    },
  …
});
```

Reusing our old code

Looking back at *Chapter 1, Your First Mapping Application*, we were able to check out the map, click on it, and get some results. Wouldn't it be great if we could reuse some of that code? Well, with a few changes, we can reuse a lot of our code.

The first thing we need to tackle is assigning our map click event. Since we don't get to do it all the time, the most logical time to assign the map click event is when our dijit is visible. We'll modify our show and hide functions so that they assign and remove the click handler, respectively. We'll name the map-click event handler `_onMapClick()`. We're also going to load the module `dojo/_base/lang`, which helps us with objects. We'll use the `lang.hitch()` function to reassign a function's `this` statement:

```
define([…, "dojo/on", "dojo/_base/lang", …],
function (…, dojoOn, lang, …) {
  …
    show: function () {
      domStyle.set(this.domNode, "display", "block");
      this._mapClickHandler = this.map.on("click", lang.hitch(this,
      this._onMapClick));
    },
```

```
hide: function () {
  domStyle.set(this.domNode, "display", "none");
  if (this._mapClickHandler && this._mapClickHandler.remove) {
    this._mapClickHandler.remove();
  }
},

_mapOnClick: function () {}
...
```

As a side note, while JavaScript doesn't support private variables, the typical naming convention for JavaScript objects states that, if a property or method starts with an underscore (_) character, it's considered private by the developer.

In our _onMapClick() method, we'll reuse the click event code from *Chapter 1, Your First Mapping Application*, with a few notable exceptions. Remember now that we're not referencing the map as a variable, but as a property of the widget. The same thing is true for the IdentifyTask, and any other methods we may call in this dijit. To refer to properties and methods within a method, the variable must be preceded by this. In this case, this will refer to the widget, which we made sure of when we used the dojo/_base/lang library on the _onMapClick() call. If your application fails at the map click event, it's probably because you didn't properly assign the variable with the correct this context:

```
define([…, "esri/tasks/IdentifyParameters", …],
function (…,IdentifyParameters, …) {
  ...
  _onMapClick: function (event) {
    var params = new IdentifyParameters(),
      defResults;

    params.geometry = event.mapPoint;
    params.layerOption = IdentifyParameters.LAYER_OPTION_ALL;
    params.mapExtent = this.map.extent;
    params.returnGeometry = true;
    params.width = this.map.width;
    params.height= this.map.height;
    params.spatialReference = this.map.spatialReference;
    params.tolerance = 3;

    this.map.graphics.clear();
    defResults =
    this.identifyTask.execute(params).addCallback
    (lang.hitch(this, this._onIdentifyComplete));
    this.map.infoWindow.setFeatures([defResults]);
    this.map.infoWindow.show(event.mapPoint);
  },
  ...
```

Loading more templates

The Y2K society had requested that we display the popups as something more organized and logical. In our web developer minds, nothing screams organized data more than a table. We can set up and organize tables for our templates, but creating those long strings to organize a long list of results would be daunting. Having HTML strings in our JavaScript code would be a pain to edit, since most IDEs and syntax-highlighting text editors don't highlight HTML in our JavaScript.

But then we remember that we can load HTML into our JavaScript using the dojo/text module. If we define the content of our InfoTemplates using little HTML snippets, and load them using dojo/text, the process will be much more streamlined:

```
define([…,
  "dojo/_base/array",
  "esri/InfoTemplate",
  "dojo/text!./templates/StateCensus.html",
  "dojo/text!./templates/CountyCensus.html",
  "dojo/text!./templates/BlockGroupCensus.html",
  "dojo/text!./templates/BlockCensus.html"
],
function (…,
  arrayUtils, InfoTemplate, StateTemplate, CountyTemplate,
  BlockGroupTemplate, BlockTemplate) {

  _onIdentifyComplete: function (results) {

    return arrayUtils.map(results, function (result) {
      var feature = result.feature,
        title = result.layerName,
        content;

      switch(title) {
        case "Census Block Points":
          content = BlockTemplate;
          break;
        case "Census Block Group":
          content = BlockGroupTemplate;
          break;
        case "Detailed Counties":
          content = CountyTemplate;
          break;
        case "states":
          content = StateTemplate;
          break;
        default:
```

```
            content = "${*}";
        }

        feature.infoTemplate = new InfoTemplate(title, content);
        return feature;
    });
  }
 });
});
```

I'll leave the contents of the HTML `infoTemplates` up to you as an exercise.

Back to our app.js

We didn't forget about our `app.js` file. We stubbed out the code for loading the map and the `Census` dijit, but we didn't assign anything else. We need to assign a click event handler for `Census` button to toggle the visibility of the `Census` widget. For that, we'll use the `dijit/registry` module to load the `dijit` button from its id. We'll load the map and the `mapService` for the `Census` dijit, and add an event listener to the `Census` button click event that will show the `Census` widget. We'll use the `dojo/_base/lang` module again to make sure the `Census.show()` function is properly applied to the `Census` dijit:

```
require([
  …
  "dojo/_base/lang",
  "dijit/registry",
  …
], function(
  … lang, registry, …
) {
  …
  var census = new Census({
    map: map,
    mapService:
    "http://sampleserver6.arcgisonline.com/arcgis/rest/services/
    Census/MapServer/"
  }, "census-widget");

  var censusBtn = registry.byId("census-btn");

  censusBtn.on("click", lang.hitch(census, census.show));

});
```

When we run our code, it should look something like the following:

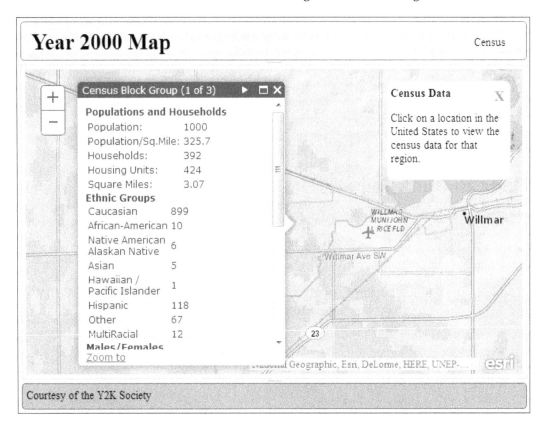

Summary

Over the course of this chapter, we have learned about many of the features offered by the Dojo framework, which is packaged with the ArcGIS API for JavaScript. We've learned about the three major packages in Dojo, and how to use them to create applications. We learned how to create our own modules, and we modified an existing application to make a module that could be imported into any application.

In the next chapter, we will learn how the ArcGIS API for JavaScript communicates with ArcGIS Server through REST services.

4
Finding Peace in REST

We've spent a few chapters discussing and building applications using the ArcGIS API for JavaScript. We've used different API tools to communicate with ArcGIS Server about its map services. But how does the API communicate with ArcGIS Server?

In this chapter, we'll focus on ArcGIS Server. We'll look at how it implements a REST interface. We'll review the ArcGIS REST API, outlined at `http://resources.arcgis.com/en/help/arcgis-rest-api/#/The_ArcGIS_REST_API/02r300000054000000/`. This describes the file structure and the format of the data that passes between the server and the browser. Finally, we'll extend the application from the previous chapter by changing the popup highlighted symbols.

What is REST? REST stands for **Representational State Transfer**. It's a software architecture that focuses on the interface between server and client using a hypermedia environment. It limits the actions that can be performed between the client and the server, but provides enough user information and documentation for the user to navigate amongst data and states.

In our discussion of ArcGIS Server and REST in this chapter, we'll cover the following topics:

- Handling REST endpoints and data formats
- The hierarchy of ESRI REST Services, as seen through a browser
- Common REST data in JSON format
- How to use ArcGIS REST Services and JSON as defined in the REST API Requirements for a REST service

REST is a methodology based around web pages. A website presents a state (the URL), data transfer (HTML page, CSS, and JavaScript), and a well-documented way to navigate between the different states (links). While understanding a website to be RESTful is well and good, what makes ArcGIS Server so RESTful?

In order for a web service to be considered RESTful, it must meet the following requirements:

- **Client-Server**: The roles of the client and server are clearly defined. The client doesn't care if the server contains one or one million records, and the server does not depend on a particular UI for the client. As long as the interface between the client and server remains the same, the client and server code can be changed independently.

- **Stateless**: The client handles the state of the application, whereas the server does not have to keep up with it. The client has all it needs to make a request, including the necessary parameters and security tokens.

- **Cacheable**: Sometimes client applications cache data for faster performance because the World Wide Web delivers data asynchronously. The server needs to tell the client which requests can be cached and for how long.

- **Layered system**: The server-side application can be placed behind a load balancer, a secure system, or a proxy, with no noticeable effect on the client-side application.

- **Code on demand (optional)**: The server can provide code for the client to run. Examples include Java applets or JavaScript scripts. Not all REST services do this.

- **Uniform interface**: With a REST service, the server provides a uniform interface through which the client can interact with the data. The uniform interface can be broken down further into four principles.

  - Information requests that include the identification of the resource. This includes everything from the data source to the output type.

  - The client has enough information from a request to manipulate or delete data.

  - Messages from the server contain instructions about how to use them.

  - A state is handled by the client using hypermedia (web pages, query parameters, and session states).

If you look into the ArcGIS Server implementation, you'll see that it meets these criteria. Therefore, it's considered RESTful.

Looking at a map server

ArcGIS Server provides web access to its map service contents. To access the content, you need to know the ArcGIS Server name and the site name. By default, ArcGIS Server is reached through port 6080. It can also be reached through port 80 if it has been configured and licensed to deliver web content. REST Service endpoints can be reached through your browser with this address:

```
http://<GIS Server Name>:6080/<site name>/rest/services
```

Where GIS Server Name refers to the ArcGIS Server machine, and site name refers to the ArcGIS Server instance which, by default, is arcgis, the port number is optional if ArcGIS Server has been set up to deliver traffic on port 80. This is the default port for Internet traffic.

When there are multiple GIS Servers, often handling a large load of traffic or complicated services, a web adaptor may be installed. The web adaptor routes traffic to multiple ArcGIS Servers, based on service requests, load balancing, and other related issues. The web adaptor also provides a layer of security, whereby ArcGIS Server machine names are not directly exposed to the outside world. To access the REST service through the web adaptor, use the following URL.

```
http://<web server name>/<web adaptor name>/rest/services
```

As long as ArcGIS Server is accessible from our computer, we can access information in the web browser. By default, the service data is presented as HTML. From there we can see the properties of the REST service, and follow links to other services and actions on the server. This lets developers review and test map services, independent of any application they create.

ArcGIS REST services provide a great way to view and interact with service data using HTML, which is good for presentation. Almost all of our applications will interact with the REST service through server requests. ArcGIS Server can therefore communicate through REST using another data format called JSON.

Working with JSON

JavaScript Object Notation (JSON) provides a structured data format for loosely defined data structures. A JSON object is built with other JSON objects, including strings, numbers, other objects, and arrays. Any data is allowed, as long as everything is self-contained and there is no sloppy formatting with missing brackets and braces.

There are a number of ways to test for valid JSON. Visit `http://jsonlint.com`, where you can copy and paste your JSON and submit it for validation. It will point out missing or broken formatting issues, as well as how to resolve them.

As you read through the examples in this book you'll see that JSON isn't always formatted in the same way, especially with JSON object key fields (or property names). JSON validators require that all string items are enclosed in quotes. Single or double quotes will work, as long as you put the same marks at the end of a string as those at the beginning. This includes both JSON object key fields. JavaScript interpreters in a browser are more flexible, key fields do not have to be enclosed by quotes. It all depends on how you're testing the JSON.

Before JSON was developed, data was passed from server to client in a format called **Extensible Markup Language (XML)**. XML is a document markup language that shows data in a format both humans and machines can read. The XML format can be read and parsed by a number of programming languages.

There are two main reasons why JSON is the preferred data format for web applications when compared to XML. First, JSON data can be consumed immediately by JavaScript applications. XML requires extra steps to parse the data into a usable object. Secondly, JSON data takes up less space. Let's explore that by comparing two snippets of data. The following snippet is written in XML:

```
<mountain>
  <name>Mount Everest</name>
  <elevation>29029</elevation>
  <elevationUnit>ft</elevationUnit>
  <mountainRange>Himalaya</mountainRange>
  <dateOfFirstAscent>May 29, 1953</dateOfFirstAscent>
  <ascendedBy>
    <person>
      <firstName>Tenzing</firstName>
      <lastName>Norgay</lastName>
    </person>
    <person>
      <firstName>Edmund</firstName>
      <lastName>Hillary</lastName>
    </person>
  </ascendedBy>
</mountain>
```

Now, here's the same data written in JSON:

```
{
    "type": "mountain",
    "name": "Mount Everest",
    "elevation": 29029,
    "elevationUnit": "ft",
    "mountainRange": "Himilaya",
    "dateOfFirstAscent": "May 29, 1953",
    "ascendedBy": [
        {
            "type": "person",
            "firstName": "Tenzing",
            "lastName": "Norgay"
        },
        {
            "type": "person",
            "firstName": "Edmund",
            "lastName": "Hillary"
        }
    ]
}
```

The same data in JSON counts 62 characters less than that in XML. If we take out the line breaks and extra spaces, or **minimize** the data, the JSON data is 93 characters shorter than the minimized XML. With bandwidth at a premium, especially for mobile browsers, you can see why JSON is the preferred format for data transmission.

JSON and PJSON formatting

JSON comes in two flavors. The default JSON is minimized, with all the extra spaces and line returns removed. Pretty JSON, or PJSON for short, contains line breaks and spacing to show the structure and hierarchy of the data. The previous Mount Everest example shows what PJSON looks like. While PJSON is easier to read, and therefore easier to troubleshoot for errors, the minimized JSON is much smaller. In the example, the PJSON has 397 characters, while the minimized version has only 277 characters, a 30 percent decrease in size.

When viewing ArcGIS REST service data, you can change the format of the data by adding an f query parameter to the REST Service URL. It should look like the following URL:

```
http://<GIS web service>/arcgis/rest/services/?f=<format>
```

Here, you can set f=JSON to receive the raw JSON data, or f=PJSON to receive the human-readable pretty JSON (or padded JSON, if you prefer). Some browsers, such as Google Chrome and Mozilla Firefox, offer third party extensions that reformat raw JSON data into PJSON without making the request.

Service levels

Let's start by viewing the sample ArcGIS Server services at http://sampleserver6. arcgisonline.com/arcgis/rest/services. When we request the page as HTML, we notice a few things. First, the version of ArcGIS Server is shown (version 10.21 at the time of writing). The version number is important because many features and bits of information may not be present in older versions. Secondly, we see a list of links pointing to folders. These are map services grouped in any way the publisher chooses. We also see a list of map service links below the folder lists. Finally, at the bottom of the page, we see supported interfaces. In this site, we can see the REST interface that we're familiar with. The other interfaces will not be covered in this book. Here's a picture of the service:

If we change the format of the REST Service request in our browser to Pretty JSON, by adding `?f=pjson` to the end of the URL, we can see roughly how the ArcGIS JavaScript API would see this location:

```
{
  "currentVersion": 10.21,
  "folders": [
   "Elevation",
   "Energy",
   "LocalGovernment",
   "Locators",
   "NetworkAnalysis",
   "Sync",
   "Utilities"
  ],
  "services": [
   {
    "name": "911CallsHotspot",
    "type": "GPServer"
   },
   {
    "name": "911CallsHotspot",
    "type": "MapServer"
   },
   {
    "name": "Census",
    "type": "MapServer"
   },
   {
    "name": "CharlotteLAS",
    "type": "ImageServer"
   },
   {
    "name": "CommercialDamageAssessment",
    "type": "FeatureServer"
   },
   {
    "name": "CommercialDamageAssessment",
    "type": "MapServer"
   },
   {
```

Here, the JSON object returned includes the numeric `currentVersion`, an array of folder names, and an array of services objects. The service JSON objects contain a name and a type attribute, which tells you what kind of service you're dealing with, and gives you the components you need to construct the URL link to those services. This format is as follows:

```
http://<server>/arcgis/rest/services/<service.name>/<service.type>
```

If we follow the link to our census map service, we can see more details.

Map services

A map service gives applications access to map data published with ArcGIS Server. It contains information about the map layout, format, contents, and other items necessary to properly render the map with the various ArcGIS API's. The map service URL is formatted as follows:

```
http://<ArcGIS Server REST Services>/<mapName>/MapServer
or
http://<ArcGIS Server REST Services>/<folder>/<mapName>/MapServer
```

When you navigate to a map service using your browser, you're presented with a lot of information about the map service. The HTML provides links to view the data in different applications, including the ArcGIS JavaScript API and ArcMap. Google Earth is also available if the map service is published to serve data in that format. The HTML for the map service also provides a lot of metadata to help you understand what it's offering. These properties include the **Description**, **Service Description**, **Copyright Text**, and the **Document Info**.

Some of the map service properties can be difficult to understand without some context. We'll review some of the important ones. Remember that properties in this list show how they are listed in the HTML. When shown in JSON, these items are camel-cased (first letter lowercase, no spaces, and capital letters to start each new word after the first).

- **Spatial reference**: How the layout of the map compares with the real world, which we'll discuss a little later.

- **Single fused map cache**: Lets you know whether the map data has been cached, or if it is dynamic. You can load the layer by using either ArcGISTiledMapServiceLayer or ArcGISDynamicMapServiceLayer, respectively.

- **Initial extent/full extent**: When you first load the map with the ArcGIS JavaScript API, the initial extent describes the bounding box of the area you see the first time. The full extent is the expected full area of the map service, which may be much wider than all the data.

- **Supported image format types**: When ArcGIS Server draws the map layers as tiles, these are the image formats that can be returned. PNG32 is recommended if your data has a lot of semi-transparencies and colors, while PNG8 works well with very simple symbols.

- **Supports dynamic layers**: If true, the developer can change the symbology and layer definitions when displaying the map service.

- **Max record count**: When submitting a query, identify or some other search, this is the maximum number of results that can be returned by the map service. This information can only be changed by server-side changes to the map service.

Finally, the Map Service HTML provides links to a number of related REST Service endpoints. Most of these links extend the existing URL and provide more information about the map service. As a bare minimum, the following should be present:

- **Legend**: Displays the symbology of the layers in the map service.

- **Export map**: This feature lets you download an image showing an area of the map that fits within a specific bounding box. You can specify parameters.

- **Identify**: This lets you identify features within all layers of a map service, based on the geometry passed in. This functionality is used by `IdentifyTask`.

- **Find**: This lets the user search for features based on the presence of a line of text passed to it. This functionality is implemented by `FindTask`.

Map service layers

When exploring the layers of a map service, it helps to know what to look for. Map services list the basic contents of their layers within an array of objects in their layer properties.

All layer objects have the same format, with the same properties. Each layer object has a numeric `id` property that refers to the layer's zero-based position in the list. Layer objects also have a `name` property that comes from how the layer was named in the map service. These layers also have a `minScale` and `maxScale` property, showing the range within which the layer is visible (with a 0 value meaning there is no `minScale` or `maxScale` limitation). When determining visibility, the layer object also contains a Boolean `defaultVisibility` property that describes whether the layer is initially visible when the map service loads.

Map service layer objects also contain information about their layer hierarchy. Each map layer object contains a parentLayerId and a subLayerIds property. The parentLayerId is a number that refers to the index of the parent group layer for the specific layer. A parent layer id of -1 means the layer in question has no parent layer. The subLayerIds are an integer array of the indexes where you can find the sublayers for the particular parent layer. If a layer has no sublayers, the subLayerIds will be a null value, instead of an empty list. You can see an example of map service layers in the following code:

```
layers: [
  {
    "id" : 0,
    "name" : "Pet Lovers",
    "parentLayerId" : -1,
    "defaultVisibility" : true,
    "subLayerIds" : [1, 2],
    "minScale" : 16000
    "maxScale" : 0
  },
  {
    "id" : 1,
    "name" : "Dog Lovers",
    "parentLayerId" : 0,
    "defaultVisibility" : true,
    "subLayerIds" : null,
    "minScale" : 16000
    "maxScale" : 0
  },
  {
    "id" : 2,
    "name" : "Cat Lovers",
    "parentLayerId" : 0,
    "defaultVisibility" : true,
    "subLayerIds" : null,
    "minScale" : 16000
    "maxScale" : 0
  }
],

    ...
```

In the preceding snippet, the map service has three layers. The Pet Lovers layer is actually a parentLayer, and corresponds to a group layer assigned in an ArcMap .mxd file. There are two layers in parentLayer: Dog Lovers and Cat Lovers. All layers are visible by default, and the layers do not appear until the map is at a scale lower than 1:16,000, according to minScale. The maxScale property is set to zero, meaning there is no maximum scale where the layer turns off again.

Feature services

Feature services are similar to map services, but provide more functionality. Their content can be edited, if the database and map settings support those operations. They display their feature symbology without the need of a legend service. Their symbology can also be modified client-side, by changing their renderer. The URL of a feature service is similar to a map service, except that it ends with FeatureServer, as shown in the following:

```
http://<GIS-web-
    server>/arcgis/rest/services/<folder>/<mapname>/FeatureServer
```

The feature service differs first and foremost in its capabilities. Apart from allowing you to query data, feature service capabilities allow the user to create, update, and/or delete records. Those familiar with CRUD operations will recognize those words as the C, U, and D in CRUD (the R stands for read, which is what happens when you query for results). The capabilities include editing if create, update, or delete are allowed. Also, if the feature service supports file attachments to data, such as photos, the capabilities will include the word "upload".

There are other feature service properties that may help you learn more about the service. They include the following:

- **Has Versioned Data**: Lets you know that the geodatabase has versioning enabled, which allows edits to be undone/redone.

- **Supports Disconnected Editing**: Data can be checked out and edited in an environment without an Internet connection. When the application connects to the Internet again, the data can be checked back in.

- **Sync Enabled**: If this is true, feature data can be synced between the geodatabase the data comes from, and another geodatabase (a topic for another book).

- **Allow Geometry Updates**: If editing is allowed, this lets the API know if the feature geometries can be edited or not. Due to certain permissions, the application might only allow for updates to the feature attributes, while the geometries remain unchanged.

- **Enable Z Defaults**: If the data contains height data (z), default values are assigned in the map service.

Layer level

Map services and feature services are made up of layers. These layers group together geographic features with the same geometry type and the same sets of properties. Layers are referred to by their numerical index in the list. The layer index starts at 0 for the bottom layer, and goes up one for each additional layer. The URL might look something like this for the first layer in a map service:

```
http://<GIS-web-
    server>/arcgis/rest/services/<folder>/<mapname>/MapServer/0
```

Map layers offer a whole host of data to help you understand what you're viewing. The layer's `name` property comes either from its name in the `.mxd` file, or from the layer in the Table of contents, if the file is unsaved. The map layer also provides a description, and copyright data. The display field property tells the map service what to use when labeling features, if labeling is turned on.

Map layers also provide important data that you can use in your application. The `type` parameter tells you the geometry of the layer, whether it's a point, line, or polygon. Default visibility lets you know if the layer was originally visible or not when the map service began. Minimum scale and maximum scale affect visibility, depending on your zoom level. The map service also lets you know if the layers have attachments, can be modified with different renderers, and how many results can be returned from a query.

Fields

A map service layer provides information about its attribute by means of the field property. The field property is an array of field objects with similar formats. All fields have a type, a name, and an alias attribute. The type refers to the data type of the field, whether it's a string or an integer, or something else. A current list of supported types can be found at `http://resources.arcgis.com/en/help/arcgis-rest-api/#/field/02r300000051000000/`. The name attribute is the field name for the property, as found in the geodatabase. Field names don't contain spaces or special characters.

The `alias` field is a string that shows the field `name` for presentation purposes. Unlike the `field` name, the `alias` can have spaces or other special characters. If no `alias` is assigned in the geodatabase or the map service, the `alias` field is the same as the field `name`. For instance, when creating the map service with ArcMap, you might have some data for a block with a field name `NUMB_HSES`. If you want to show the values for this property in a chart, the field name may look rough and a little confusing. You can then add an alias to the `NUMB_HSES` field by calling it `Number of Houses`. That `alias` provides a much better description for the field:

```
{
  "type": "esriFieldTypeInteger",
  "name" "NUMB_HSES",
  "alias": "Number of Houses"
}
```

Domains

Field objects may also have domain attributes assigned to them. Domains are limitations on field values imposed at the geodatabase level. Domains are uniquely created in the geodatabase, and can be assigned to feature classes and table fields. Domains make it easier to input the correct values by restricting what can be entered. Instead of allowing users to mistype street names in a report service, for instance, you might provide a field with a domain containing all the correctly typed street names. The user can then select from the list, rather than have to guess how to spell the street name.

The ArcGIS REST API supports two varieties of domains: ranges and coded values. Ranges, as the name implies, set a minimum and maximum numeric value for a feature attribute. One example of a range might be an average user rating for restaurants. The restaurant might get somewhere between one and five stars, so you wouldn't want a restaurant to accidently get a value of 6 or a value of less than 1. You can see an example of a rating field with that range domain in this snippet:

```
{
  "type": "esriFieldTypeInteger",
  "name": "RATING",
  "alias": "Rating",
  "domain": {
    "type": "range",
    "name": "Star Rating",
    "range": [1, 5]
  }
}
```

A coded value domain provides a list of code and value pairs to use as legitimate property values. The coded value list contains items with a name and a code. The code is the value stored in the geodatabase. The name is the text representation of the coded value. They're useful in that users are forced to select a valid value, instead of mistyping a correct value.

In the following example, we can see a field with a coded value domain. The field contains state abbreviations, but the domain allows the user to see entire state names:

```
{
    "type": "esriFieldTypeString",
    "name": "STATE",
    "alias": "State",
    "length": 2,
    "domain": {
        "type": "codedValue",
        "name": "State Abbreviation Codes",
        "codedValues": [
            {"name": "Alabama", "code": "AL"},
            {"name": "Alaska", "code": "AK"},
            {"name": "Wisconsin", "code": "WI"},
            {"name": "Wyoming", "code": "WY"}
        ]
    }
}
```

In the preceding example, state names are stored in two letter code form. The domain provides a full name reference table with the full names of the states. If you were to send queries for features using this field, you would use the code values. Querying for all features where STATE = 'Alaska' would yield no results, while a query where STATE = 'AK' may give you results.

 Note that the code and the value don't have to be of the same type. You can have numeric codes for, say, water line part numbers, and coded values to show their descriptive names.

Related tables

Tables with non-geographic data can be published in a map service. These tables may provide data related to map features, such as comments on campground locations or the sales history of a property. These tables can be searched and queried like features. Relationships between map layers and tables can also be published and searched.

Layers and tables can be joined, either by using geodatabase relationship classes, or ad-hoc relationship assignments in ArcMap. When published with ArcMap, those relationships are preserved in ArcGIS Server. The connection between related features and tables is stored within the `relationships` property of the layer and table. A developer can query related data, based on a selection in the parent feature class.

Relationship objects have the same general format. Each relationship object contains a numerical `id` and `relatedTableId`. The `relatedTableId` is linked to the `RelationshipQuery` object to query for related results. The role describes whether the current layer or table is the origin or the destination of the relationship. Cardinality describes whether a single origin object has one or several destinations related to it.

> When querying for results, results return much faster if you start with the origin and use `RelationshipQuery` on the destination tables. Starting with the destination tables may take significantly longer.

Common JSON objects

The ArcGIS REST API defines the formats for JSON data objects commonly used by the system. The ArcGIS API for JavaScript uses this format to communicate with the server. When you look at the network traffic in your browser's developer console, you'll see these common JSON objects in both the requests and responses. Let's look more closely at two of these JSON object definitions: geometries and symbols.

Geometries

Geometry JSON is a common data format used for ArcGIS Server processes. It makes sense, since finding geographic locations relies on geometry. Geometries come in a variety of shapes and sizes.

Spatial reference

As mentioned in previous chapters, spatial reference refers to the calculations that represent the earth on a map, either as a spheroid (almost a sphere), or as a 2-dimensional representation of a land surface. Instead of documenting all the possible factors that go into calculating a spatial reference, a general format is followed, either by using an assigned **well-known ID (WKID)**, or **well-known text (WKT)**.

For spatial references with well-known ids, the spatial reference JSON is made up of one required parameter (WKID), and three optional parameters. The WKID is a numerical reference to the spatial reference, which can be searched at http://www. spatialreference.org. Common values include `wkid 4326`, which stands for WGS84, and is typically used for latitude and longitude. Optional spatial reference object parameters include the latest WKID or `latestWkid`, a vertical coordinate system WKID or `vcs Wkid`, and the latest vertical coordinate system WKID or `latest vcs Wkid`. The last two are used for features with three-dimensional points, including height.

Spatial references can also be defined using strings, in the case of the well-known text. The well-known text details the necessary items used to flatten the map mathematically.

We will not be calculating anything with spatial references. For our applications, we'll use them either for data comparison, or as parameters for service requests. Let's say we need the latitude and longitude coordinates for the results of a query, but the data isn't in a decimal degree latitude/longitude spatial reference. We can query the data, but set the `query.outSpatialReference` to `wkid 4326`, and ArcGIS Server performs the necessary calculations to give us the results in the format we need.

Points

Points are the simplest geometry, and therefore the simplest to explain. They have three required parameters, an x value, a y value, and a `spatialReference` value. The x value refers to the longitude, or easting value, depending on the coordinate system. The y value refers to the latitude or the northing value. Points may also have an optional z value to describe the height, and an optional m value to describe the slope. You can see an example of a point REST object in the following:

```
// a two-dimensional point
{x: -95.25, y: 38.09, spatialReference: {wkid: 4326}}

// a three-dimensional point
{x: -38.93, y: -45.08, z: 28.9, spatialReference: {wkid: 4326}}
```

In the preceding example, the point has an x and y value, plus a spatial reference. As we learned in the previous section, a well-known id of 4326 means the x and y values are longitude and latitude. In the second example, a z value was added, indicating a height.

Multipoints

Multipoints, as you may recall from previous chapters, define a cluster of points.
The points share the same spatial reference (`spatialReference`). Multipoints define
points by using a `points` property, which is a two dimensional array of points. The
lowest array contains the x, y, and possibly z (height) and m (slope) values. If the
points contain z and/or m values, the multipoint JSON will also have true values for
its `hasZ` and `hasM` properties:

```
// a two-dimensional multipoint
{
  "points": [[12.831, 48.132], [19.813, 49.908], [-90.10,
  83.132]],
  "spatialReference": {wkid: 4326}
}
```

This example shows three points grouped together as a multipoint, with locations
defined using latitude and longitude coordinates (wkid 4326).

Polylines

Polylines can be either one line, or a group of lines, that describe the same linear
feature on a map. The polylines JSON is made up of a spatial reference and a
three-dimensional array of points in its `paths` property. The lowest level of the
polyline paths array contains the x, y, and possibly z and m coordinates for a point
on the line. If z and m values are present, the polyline JSON object has its `hasZ` and
`hasM` values set to true, respectively.

A polyline may contain more than one line. For instance, if part of a railroad line is
abandoned, the railroad still owns the lines on either side of that section. The two
good sections would still be considered a polyline. This is why the polyline contains
three levels of arrays. The preceding code symbolizes a single line with a string of
latitudinal and longitudinal coordinates:

```
// a two dimensional path.
{
  "paths": [[[43.234,-28.093], [44.234,-32.232], [43.239,-33.298],
  [49.802,-35.099]]],
  "spatialReference": {wkid: 4326}
}
```

Polygons

Polygons provide a solid shape to areas. A polygon JSON is made up of a spatial reference and a `rings` parameter. The rings parameter contains a triple array with one or more lists of points, where the last point is the same as the first point. In this way, the polygon closes in on itself and provides an interior and an exterior. The points within the polygon rings can contain `z` values for height. In that case, the polygon would have the optional `hasZ` property set to true.

A polygon JSON can contain more than one closed loop of points, and these polygons are called **multipart polygons**. An example of this would be a plot of land that has a road built down the middle of it. If the road no longer belongs to the owner, the owner still retains ownership of the two halves of the property divided by the road. Since the land used to be as one, it is still treated as one unit, but with two parts. The following JSON code illustrates a multipart polygon. The polygon has two `rings`, each with three points:

```
// a two dimensional polygon with two rings.
{
  "rings": [[[-85.032,18.098], [-85.352,18.423], [-85.243,18.438],
  [-85.032,18.098]], [[85.042,18.098], [84.995,18.008],
  [85.123,18.900], [85.042,18.098]]],
  spatialReference: {wkid: 4326}
}
```

Envelopes

Envelopes, also known as extents, are the bounding boxes that represents the minimum and maximum x and y values for items on the map. The following example shows an area between `42.902` degrees and `53.923` degrees longitude, and `-23.180` degrees and `-18.45` degrees latitude:

```
{
  "xmin": 42.902, "xmax": 53.923, "ymin": -23.180, "ymax": -18.45,
  "spatialReference": {wkid: 4326}
}
```

Symbols

The symbol JSON contains the minimum amount of data necessary to describe the properties of the symbols. As we learned in previous chapters, a graphics symbol describes the appearance of the graphic, affecting the color, line thickness, and transparency, among other features. Different geometry types have corresponding symbol types to demonstrate their features. We're going to look at how the ArcGIS REST API describes the different symbol types, and how to use them in our application.

Color

Colors are defined in ArcGIS REST services as three to four element arrays with integer values from 0 to 255. The first three values stand for the red, green, and blue color values, respectively. Higher numbers mean lighter colors, and lower numbers mean darker colors, with [0, 0, 0] representing the color black.

The fourth value refers to the alpha, or the opacity of the color. A value of 255 is completely opaque while a value of 0 is completely transparent, and a value somewhere in between is somewhat see-through. If the fourth value is left out, it is considered to be completely opaque (255). This description differs from the esri/Color description of opacity, which defines the value as a decimal value between 0.0 and 1.0.

Simple line symbol

As for symbols, we start with the simple line symbol because it is used by all the other symbols. The line consists of four basic properties: a type, a style, a width, and a color. The type, esriSLS, is always the same for a simple line symbol. The width describes a pixel width for the line, and color describes the numeric RGB value described in the previous color section. style is one of several string constants prefixed by esriSLS. The choices include Solid, Dash, Dot, DashDot, DashDashDot, and Null. You can see an example of a JSON simple line symbol definition in the following snippet:

```
{
    type: "esriSLS",
    style: "esriSLSDash",
    width: 3,
    color: [123,98,74,255]
}
```

Simple marker symbol

The simple marker symbol describes the style of a simple point and is made up of the following properties: a `type`, `style`, `color`, `size`, `angle`, `xoffset`, `yoffset`, and `outline`. The `type` is always `esriSMS` for these symbols. Styles are defined in a similar way to the simple line symbol, except that all styles are prefixed with `esriSMS`. The style can be chosen from a `Circle`, `Cross`, `Diamond`, `Square`, `X`, or `Triangle`. `Angle` defines the rotation of the symbol, while `xoffset` and `yoffset` describe how far the drawn point is away from the real point. `Size` is numeric, and measures the pixel size of the graphic. Finally, `outline` accepts a simple line symbol definition to describe the border of the point graphic:

```
{
  type: "esriSMS",
  style: "esriSMSCircle",
  color: [255,255,255,50],
  size: 10,
  angle: 0,
  xoffset: 0,
  yoffset: 0,
  outline: {
    width: 1,
    color: [0,0,0]
  }
}
```

Simple fill symbols

Simple fill symbols can be described by the following properties: `type`, `style`, `color`, and `outline`. `Type` is always `esriSFS`, and all styles start in the same way. Styles consist of `Solid`, `Null`, and various other ways to describe lines drawn within the graphic. `Outline`, like the simple marker symbol, uses the simple line symbol to describe the border of the polygon. Look at the following code for an example of a JSON description:

```
{
  type: "esriSFS",
  style: "esriSFSSolid",
  color: [123, 99, 212, 120],
  outline: {
    type: "esriSLS",
    style: "esriSLSDashDot",
    width: 3,
    color: [123, 99, 212]
  }
}
```

Picture marker symbol

The picture marker symbol describes a point symbol with a picture graphic. The symbol JSON for this includes unique properties such as `url`, `imageData`, `contentType`, `width`, and `height`, as well as properties common to the simple marker symbol (type, `angle`, `xoffset`, and `yoffset`). The URL links to the image you want to see with the features. `Width` and `height` describe how large you want the image to be in pixels. `ContentType` refers to the file extension, such as `image/png` for a `.png` file. Finally, `imageData` is populated by the `base64` string that can be translated into an image file. Refer to the following example:

```
{
  type: "esriPMS",
  url: "images/Cranston.jpg",
  imageData: "iVBORw0KGgoAAAANSUhEUgAAABoAAAAaCAYAAACpSkzOAAAAA…",
  contentType: "image/jpg",
  width: 24,
  height: 24,
  angle: 0,
  xoffset: 0,
  yoffset: 0
}
```

Picture fill symbol

The picture fill symbol can be added to a polygon to show an image tiled inside the shape. The picture fill symbol has many of the same parameters as the picture marker symbol, including `url`, `imageData`, `contentType`, `width`, `height`, `angle`, `xoffset`, and `yoffset`. It requires the line symbol JSON for the outline. It accepts its own numeric `xscale` and `yscale` parameters, which shrink or stretch the images shown within the feature. You can see an example of a picture fill symbol JSON here:

```
{
  type: "esriPFS",
  url: "images/foliage.png",
  imageData: "iVBORwoadfLKJFDSFKJLKEWIUnjnKUHWkunUWNkJNiuN…",
  contentType: "image/png",
  outline: {
    type: "esriSLS",
    style: "esriSLSSolid",
    width: 3,
    color: [16, 243, 53]
  },
  width: 32,
  height: 32,
```

```
        angle: 0,
        xoffset: 0,
        yoffset: 0,
        xscale: 1,
        yscale: 1
    }
```

Text symbol

The text symbol JSON describe the formatting used to create custom text labels on the map. The text symbol JSON looks very different to the other symbols because it is also concerned with things such as fonts, boldness, and italics. However, the text symbol does have certain properties in common with the other symbols such as color, angle, xoffset, and yoffset.

The text symbol has some unique properties that need to be addressed. Vertical and horizontal alignment position the text in relation to the point because text symbols are typically placed around label points. Halo size and color, introduced in ArcGIS Server 10.1, describe the color outline around the text, to make it more readable against a busy aerial background. The font parameters are so numerous they require their own object to describe the CSS styling of the labels. You can see an example of a text symbol here:

```
{
  "type": "esriTS",
  "color": [158,38,12,255],
  "backgroundColor": [0,0,0,0],
  "borderLineSize": 1,
  "borderLineColor": [158,38,12,255],
  "haloSize": 2,
  "haloColor": [255,255,128,255],
  "verticalAlignment": "bottom",
  "horizontalAlignment": "center",
  "rightToLeft": false,
  "angle": 0,
  "xoffset": 0,
  "yoffset": 0,
  "kerning": true,
  "font": {
    "family": "Georgia",
    "size": 14,
    "style": "normal",
    "weight": "bold",
    "decoration": "none"
  }
}
```

Uses

While you will seldom see symbol JSON passed between the server and client, all of the symbol constructors in the ArcGIS JavaScript API can accept JSON objects to define the symbols. The developer can thus stash the symbol style in a configuration file until it's needed. The developer can also construct the JSON object with a different color palette and line defining widget before it is assigned as a symbol.

Assigning symbols with JSON is much easier than the previous methods of defining a simple symbol. The older ways included symbol constructors containing a long list of arguments. Some constructors could possibly contain nested constructors. Another way of constructing symbols include creating a base symbol, and then assigning the necessary attributes with assignment methods. Let's take a look at an example of the older way:

```
require(["esri/symbols/SimpleFillSymbol",
  "esri/symbols/SimpleLineSymbol","esri/Color", …],
function (SimpleFillSymbol, SimpleLineSymbol, esriColor, …) {
  …
  var lineSymbol = new SimpleLineSymbol()
    .setColor(new esriColor([255,128, 128, 0.8]))
    .setWidth(3)
    .setStyle(SimpleLineSymbol.STYLE_DASHDOTDOT);
  var fillSymbol = new SimpleFillSymbol()
    .setOutline(lineSymbol)
    .setColor(new esriColor([64,255,64,0.4]));
  …
});
```

Here, a base symbol is constructed and additional methods are called to assign colors, line width, and line style. Compare this to the same symbol declaration using JSON:

```
require(["esri/symbols/SimpleFillSymbol", …],
function (SimpleFillSymbol, …) {
  …
  var fillSymbol = new SimpleFillSymbol({
    style: "esriSFSSolid",
    color: [64,255,64,102],
    outline: {
      color: [255,128,128,204],
      width: 3,
      style: "esriSLSDashDotDot"
    }
  });
  …
});
```

While the JSON code takes up an extra line of code, it's more compact and easier to read. Also, this symbol assignment doesn't require the client to load two extra modules just to assign a color and a line style. It's all included in a single construction call.

> When constructing symbols using JSON, you don't have to add every single JSON attribute defined in the symbol. When the symbols are constructed, the attributes of the JSON object are mixed into a default symbol. That means that, if you construct a line symbol using the following code, instead of a 1 pixel wide black line, you'll create a 1 pixel wide green line.
>
> ```
> Var symbol = new SimpleLineSymbol({color: [0,255,0]});
> ```

Back to our application

Now that we have an idea of how to work with JSON data, we can improve our Year 2000 Census map. We've come a long way in styling our application but the default highlight color for the map's `infoWindow` clashes with the rest of the site.

How about that InfoWindow

Our clients at the Y2K Society just responded with feedback on the site. They liked the color palette for the page. Unfortunately, they complained that, when they clicked on the map, the cyan-colored highlight symbol on the map's popup clashed with the site. One user even claimed it gave him a headache to stare at it too long. This problem is fixable, and there are things we can do to make the client happy.

Let's look at the application we wrote in the previous chapter. If you haven't seen the application in *Chapter 3, The Dojo Widget System*, there are code samples included with this book. We created a widget using the Dojo framework to interact with the map and reveal the census data for the year 2000. We will modify the widget so that it can modify the map properties for us.

Open the `Census.js` file in your favorite text editor. If you look it over, you'll see where we construct the Census widget, then use the `show()` and `hide()` functions to toggle map clickability. We would like to modify the symbols for the map's `infowindow` but, where do we begin?

We could change the map's `infowindow` properties in the constructor, but that could lead to problems. If you remember, in *Chapter 2, Digging into the API*, we talked about how changing map settings before it finishes loading causes an error and possibly stops the application. We'll have to check if the map is loaded before we make any changes to its `infowindow`. If it hasn't loaded, we'll assign an event listener to the map's load event. If it has loaded, we'll go ahead and run the event listener. We'll call `_onMapLoad()` event listener, as you can see in the following code:

```
...
this.map = options.map || null;
this.domNode = srcRefNode;

if (this.map.loaded) {
  this._onMapLoad();
} else {
  this.map.on("load", lang.hitch(this, this._onMapLoad));
}
...
```

A line below the `Census` dijit `constructor()` method, we'll add the new `_onMapLoad()` method. We know that the map has loaded there, and that means the `infoWindow` object should be ready as well. We'll modify the `infoWindow` object's `markerSymbol` and `fillSymbol` properties with our own symbology. Now, we haven't added modules for the `SimpleMarkerSymbol` and `SimpleFillSymbol` yet, so we'll add the references to those in our define statement:

```
define([
  ...
  "esri/InfoTemplate",
  "esri/symbols/SimpleMarkerSymbol",
  "esri/symbols/SimpleFillSymbol"
], function (

  InfoTemplate, MarkerSymbol, FillSymbol
) {
  ...
    _onMapLoad: function () {
      // change the infoWindow symbol
      this.map.infoWindow.markerSymbol = new MarkerSymbol({});

      this.map.infoWindow.fillSymbol = new FillSymbol({});
    },

});
```

So, we know that symbol constructors can accept JSON objects to create the symbols. Let's fill in the properties in the objects we'll pass into the symbols. We'll pick some colors based on the application's color palette. We'll also adjust the fill symbol's interior transparency, so that the user can see what's underneath. Remember that, when defining a color, the four numbers represent red, green, blue, and opacity values, ranging from a full 255, to an empty 0. Also, you don't have to pass every single symbol JSON property, if you don't mind that it uses a default value instead.

```
this.map.infoWindow.markerSymbol = new MarkerSymbol({
  style: "esriSMSDiamond",
  color: [255,200,30],
  size: 12,
  outline: {
    width: 1,
    color: [111,98,34]
  }
});

this.map.infoWindow.fillSymbol = new FillSymbol({
  style: "esriSFSSolid",
  color: [255,250,169,128],
  outline: {
    width: 2,
    color: [111,98,34]
  }
});
```

We can then save and reload our application in the browser. If we've typed everything in correctly, the map should load with no errors. When we click on the **Census** button, the floating widget should load in the corner, and when we zoom in and click on an area, we should see new yellow diamonds when points are selected, and polygons highlighted in colors similar to the application. Our clients will hopefully be pleased with our color choices.

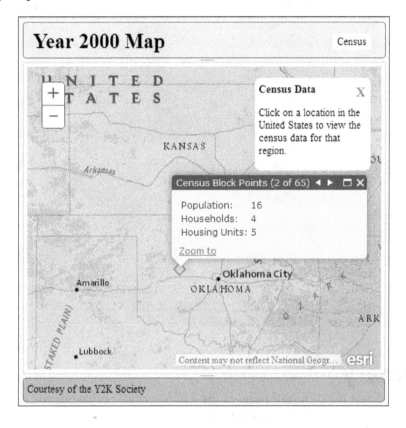

The preceding image shows the **Census Block Points** with the golden rod-colored diamond. Each selected point will show the same graphic. You can see a picture of one of the states in the following. Note that the fill color is semi-transparent, allowing the user to read the content underneath:

Summary

We have explored the different parts of the REST Service endpoints for ArcGIS Server, the primary data source for our ArcGIS JavaScript API-based applications. We've learned what it means for a service to be RESTful, and how that applies to ArcGIS Server. We've explored the organization of ArcGIS Server map services. We've learnt about what information is available when viewing REST Services in a web browser, and with JSON requests. We learned about the ArcGIS REST API, how data is formatted for requests, and how it is consumed, both by ArcGIS Server and by the ArcGIS JavaScript API. Finally, we applied some of our knowledge to improve our application.

With your understanding of how data is handled client-side and server-side, you will be able to implement many of the powerful features that ArcGIS Server offers for web-based applications. In the next chapter, we'll look into one of those powerful features: editing map data.

5
Editing Map Data

Data has to come from somewhere. In the multi-billion dollar geospatial industry, collecting data is expensive. Features visible from aerial photography need to be plotted, and features not so visible on a workstation need their GPS coordinates collected in the field. Your time is valuable, and data collection won't happen on its own.

But what if you could get others to do the work for you? What if you could create a website that let other people collect the information? Trained workers could document utility lines, or concerned citizens could report problem locations in town. By using volunteer data collection, you can quickly collect the data you need.

ArcGIS Server provides not only data visualization on a map, but editing capabilities as well. Services can be created, and applications can be built around them, which allow users to add items to a map, change their shape, edit their attributes, and even delete them. ArcGIS Server also gives the creator of the services control over which of those data changes are allowed.

In this chapter, we're going to do the following:

- Look at the data editing capabilities ArcGIS Server provides with the ArcGIS JavaScript API
- Learn about the editing controls in the API, and how they create a helpful editing experience
- Create an editing application that uses the tools in the ArcGIS JavaScript API

The use cases for webmap editing

A GIS professional doesn't have to edit all the map data on his own. Trained staff and eager volunteers can assist with data collection and map editing projects that interest them. As the developer, it's up to you to give them the tools they need to collect and update the data. The following are examples of applications which use web map editing that you can create :

- Field crews updating utility data
- Public service requests and incident reports
- Parcel classification reassignments after analysis
- Volunteer geographic information data collection

Map editing requirements

Editing geographic data using the ArcGIS JavaScript API requires some setup on ArcGIS Server. An editable **feature service** must be published on ArcGIS Server, which requires an ArcSDE geodatabase. File geodatabases, personal geodatabases, and shapefiles cannot be used to store editable data. ArcGIS Online allows you to upload editable data to ESRI's cloud service, but the data upload and editing process has requirements which are covered in *Chapter 11, The Future of ArcGIS Development*.

There are a few requirements for setting up an editable map application using ArcGIS Server and its JavaScript API. The geodatabase storing the data should be versioned, if you want to review the data before committing it to your default database. Versioned data also supports undo and redo operations. You may want to publish a read-only map service along with your editable feature service. Finally, some editing operations require a geometry service to handle geometry changes, as you cut, merge, and trim features.

Feature services

A feature service provides a web interface between data stored on the server and an application on the browser created to use it. They can be accessed through URL endpoints similar to map services. However, they produce very different results. They can be loaded on a map and queried much like dynamic or tiled services, but there is more. Feature services return graphics instead of tiles. These graphics can be queried, and even edited, if the service allows.

Feature templates

With ArcGIS Server 10.1, feature services can be published with the added functionality of **feature templates**. Feature templates give the user preconfigured features to add to the map. Feature templates are created in ArcMap, and define the symbology and predefined attributes. These templates make it easier to edit service data.

One example of Feature templates can be found on an animal sighting map. The points on the map designate where animal sightings take place. Feature templates could be created to show pictures of each major type of animal (cat, dog, bird, rabbit, deer, and so on). Values in some of the fields could be defined ahead of time. For instance, you could say that all cats are warm-blooded.

How do you, as the developer, take advantage of feature templates? Apart from demonstrating what each symbol means, there are template pickers in the ArcGIS JavaScript API's tools that not only show the feature templates, but also let you click on them and add them to your map.

Feature layer

The feature layer provides access to graphics within a feature class. The user can thus both query and edit the shapes and attributes of the graphics. We reviewed their REST service profile in *Chapter 4, Finding Peace in REST*. We load feature layers in much the same way we load dynamic and tiled services. However, their options often require more parameters, due to the editable nature of the content.

Feature service modes

When initializing a feature layer from a feature service, you have a choice as to how the data is loaded. Do you want to load it all at once? Do you want to load all the features that you can see? Do you only want to load the one you've selected, and not show the rest? In the next sections, we'll review the three feature service modes used to download data to the client browser.

Snapshot mode

Sometimes, if there is not a lot of data, it's better to download it all at once. That's what snapshot mode does. Snapshot mode downloads feature data based on time definitions and definition expressions, but it is limited by the maximum download limit. The visibility of the data is then determined by time extent.

Snapshot mode is helpful if there is not a lot of data to download, or if connectivity may be an issue during use. The user can download all the feature data at once, work with it, and then save their changes when connections become favorable again.

On demand mode

Sometimes, you're only interested in downloading the data in front of you. In that case, on demand mode is the best option. On demand mode only downloads features within the map extent. They too are affected by time definitions and definition expressions. Unlike snapshot mode, data requests are made every time the map extent changes. On demand mode is the default mode for any `FeatureLayer`.

On demand mode is typically used when there is a lot of data in the feature layer, but the user is only expected to view a small portion of it. It's very good for focussed editing tasks. It's not as good for mobile applications with lots of map navigation and connectivity issues, since some graphics will fail to load.

Selection mode

Loading features by selection is more constraining because it only shows those features that have been selected. Feature selection is handled using the feature layer's `selectFeatures()` method, in a manner similar to querying from a map service layer. In this case, the graphics returned are considered "selected". Selection methods include clicking on the map and sending a query with specific parameters. This method is very helpful if there are lots of features, and you only want to download specific ones, whether it's by area or attributes.

Editing tools

The ArcGIS JavaScript API comes with a set of widgets and modules designed specifically for editing. With the editing widgets, the user can add features to the map, change their shape, edit their attributes, and even delete them, if the services allow. Let's look at some of the tools available in the API.

Why are the editing tools in your application not working? It may be the CSS. Editing widgets are created with Dojo user controls, or dijits. These controls require the Dojo stylesheets, such as `claro.css` or `nihilo.css`. Without them, buttons stop working, and other unexpected behaviors may arise.

Edit toolbar

The edit toolbar, loaded with the `esri/toolbars/edit` module, lets the user change the shape, orientation, scale, and position of graphics on a map. We discussed it in *Chapter 2, Digging into the API*, in relation to the other toolbars. Separate controls are required to save the changes made with the edit toolbar. You can see an image of a triangle selected for the edit toolbar here:

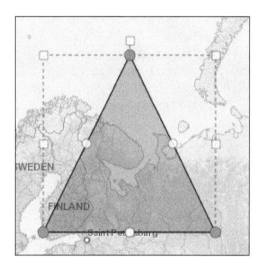

The edit toolbar requires a map in its constructor. The constructor also needs a number of optional parameters to modify its appearance and behavior. Many of the options depend on the geometry type of the data being manipulated. As of API version 3.13, here are some of the available options for the edit toolbar:

- `allowAddVertices` (`boolean`): If true, you can add vertices to a polyline or polygon.
- `allowDeleteVertices` (`boolean`): If true, you can remove vertices from a polyline or polygon.
- `ghostLineSymbol` (`line symbol`): When moving a line or polygon edge, this is the symbol for the line that shows where the new line/edge will go.
- `ghostVertexSymbol` (`marker symbol`): If you are allowed to add vertices, this is the symbol that shows where to click to insert a vertex.
- `textSymbolEditorHolder` (`string` or `HTML DOMnode`): Web page location when you want to add a text symbol editor widget
- `uniformScaling` (`boolean`): When true, resizing a polyline or polygon keeps the original ratio of width to height.
- `vertexSymbol` (`marker symbol`): When editing polylines and polygons, this is the symbol of the points at each vertex.

You can see an example of loading the edit toolbar in the following snippet:

```
require([…, "esri/toolbars/edit",
  "esri/symbols/SimpleMarkerSymbol",
  "esri/symbols/SimpleLineSymbol",  ],
function ( …, EditToolbar, MarkerSymbol, Linesymbol, …) {

  var editTB = new EditToolbar(map,… {
    allowAddVertices: true,
    allowDeleteVertices: true,
    ghostLineSymbol: new LineSymbol(…),
    ghostMarkerSymbol: new MarkerSymbol(…),
    uniformScaling: false,
    vertexSymbol: new MarkerSymbol(…)
  });

});
```

When you want to use the edit toolbar to edit a feature, you call the `activate()`
method. The `activate()` method requires two arguments, and has the option for
a third. Firstly, the method requires a tool, which is made by joining a combination
of the edit toolbar constants with the pipe | symbol. The constants include EDIT_
TEXT, EDIT_VERTICES, MOVE, ROTATE, and SCALE. Secondly, the `activate()` method
requires a graphic to edit. The final optional argument is an object similar to the one
used to create the edit toolbar. In the following code snippet, we have a graphic that
is added to the map, and a click event is assigned to it that activates the edit toolbar
to edit the graphic when it is double-clicked:

```
var editTB = new EditToolbar(…);
  …
map.graphics.add(myGraphic);
dojoOn(myGraphic, "dblclick", function () {
  editTB.activate(EditToolbar.EDIT_VERTICES | EditToolbar.MOVE |
  EditToolbar.ROTATE | EditToolbar.SCALE, myGraphic);
  dojoOn.once(myGraphic, "dblclick", function () {
    editTB.deactivate();
  });
});
```

Attribute inspector

Sometimes, you don't care where things are, you just care about the content. That's where the attribute inspector comes in. The attachment inspector provides a form with a list of editable fields and the appropriate blanks to edit them. The attachment inspector is bound to a feature layer, and displays the editable values for the selected layer. The fields in the attribute inspector respond to the field types of the attributes. Date fields show a calendar when editing. Fields with coded value domains show a drop-down list instead of a text blank. Below, you can see an example of an attribute inspector loaded in the popup, though it could be added to a separate HTML element.

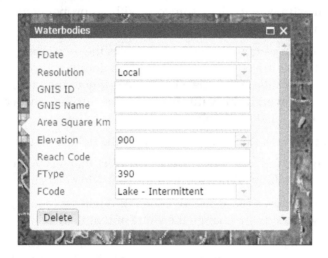

When initializing an attribute inspector, you need to define how the inspector will handle the different fields within the graphic attributes. The attribute inspector constructor accepts an `options` object, and either an HTML element or an id string reference to the element. The `options` object has one parameter called `layerInfos`, which accepts an array of `layerInfo` objects. Each `layerInfo` object contains one or more of the following parameters:

- `featureLayer` (required): The feature layer to be edited.
- `userId` (string, optional): The ArcGIS Server user id connected to the service, should the editing require token authentication. This is not needed if you have used the Identity Manager to handle logins.
- `showObjectID` (Boolean, optional): Whether you want to see the object id of the feature when it is selected. By default, this value is `false`.

- showGlobalID (Boolean, optional): Whether you want to see the global id of the feature when it is selected. By default, this value is false.

- showDeleteButton (Boolean, optional): By default, the attribute inspector shows a delete button that lets you delete the selected feature. Setting this to false removes it.

- showAttachments (Boolean, optional): When set to true, and if the feature layer has attachments, this displays an attachment editor form in the attribute inspector, which lets you view and upload files attached to the feature.

- isEditable (Boolean, optional): Lets you control whether the feature is editable. This doesn't override whether the features are editable server-side. It's just an extra way to block someone without proper credentials from editing data they shouldn't.

- fieldInfos (Objects [], optional): Gives the developer granular control over what fields are editable, and how. This does not allow the user to edit fields that aren't allowed to be edited, according to the publishing method of the feature layer. If this value is not set, the attribute inspector lists all editable fields. FieldInfo objects contain the following:
    - fieldname (string): The name of the field to be edited
    - format (object, optional): An object that lets you edit time when editing dates. When set, add the following object: {time: true}
    - isEditable (Boolean, optional): When set to false, this disables the user's ability to change the value of that field
    - stringFieldOption (string, optional): When set, the user can edit a string value either in a single-line textbox, a text area with multiple lines, or a rich-text field that includes additional formatting
    - label (string, optional): When set, this lets you override the name of the field alias from the feature service
    - tooltip (string, optional): When set, this shows a text tool tip when the user begins editing the attribute

- You can see an example of an attribute inspector being loaded with a single feature layer here:

```
var layerInfos = [{
  'featureLayer': bananaStandFL,
  'showAttachments': false,
  'isEditable': true,
  'format': {'time': true },
  'fieldInfos': [
    {'fieldName': 'address', 'isEditable':true, 'tooltip':
    'Where is it?', 'label':'Address:'},
    {'fieldName': 'time_open', 'isEditable':true,
    'tooltip': 'Time the Banana Stand opens.',
    'label':'Open:'},
    {'fieldName': 'time_closed', 'isEditable':true,
    'tooltip': 'Time the Banana Stand closes.',
    'label':'Closed:'},
    {'fieldName': 'is_money_here', 'isEditable':false,
    'label':'Is Money Here:', 'tooltip': 'There\'s money in
    the Banana Stand.'}
  ]
}];

var attInspector = new AttributeInspector({
  layerInfos: layerInfos
}, "edit-attributes-here");

attInspector.startup();
```

While the attribute inspector allows you to edit the attributes of graphics on a map, it doesn't provide an immediate way to save the edits. It is up to the developer to determine when changes to attributes are saved to the server. The developer could add a save button, or save whenever the feature is no longer selected.

Template picker

The **template picker** lets the user select from a list of feature templates to add features to the map. It displays a grid of feature templates from connected feature layers. These templates include feature names, geometry types, and preset styles. The user can click on any of the template buttons, and then draw them on the map. You can load more than one feature layer, and switch between them with ease. You can see an example in the following screenshot:

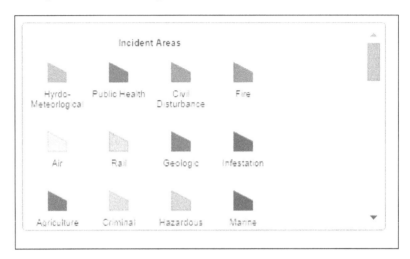

The template picker, like most dijits, requires a parameter object and, either an HTML element or a string reference to the element's id, in order to load. In the options, the template picker accepts an array of `featureLayers`. It also accepts the number of `rows` or `columns` it will create. If you don't use `featureLayers` with their own feature templates, you can define your own using configured items in the `items` array. You can also set the CSS style directly. Finally, you can control whether tooltips show when you hover over the symbols. In the following snippet, you can see an example of a template picker initialized:

```
var widget = new TemplatePicker({
  featureLayers: layers,
  rows: "auto",
  columns: 9,
  showTooltip: true,
  style: "height: 100%; width: 900px;"
}, "templatePickerDiv");
```

The preceding code shows a template picker with nine columns with tooltips to show data about the `layers` loaded in its `featureLayers` attribute. The size is 900 pixels wide, and as tall as it needs to be.

Attachment editor

There's an old saying that a picture is worth a thousand words. Sometimes, you need that picture to explain what data you're submitting. The **attachment editor** can help. The attachment editor allows the application to upload a file, usually an image, and connect it to the feature on the map. You can view other attachments, and possibly edit them or delete them, if permissions allow. Attachment editors can be loaded as part of the attribute inspector by setting the showAttachments property in the attribute editor options to true, when constructing the editor:

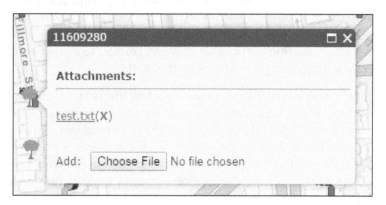

Editor dijit

The **editor dijit** provides an all-in-one editing tool to create, update, and delete map features. The editor dijit includes the template picker, attribute inspector, and an editing toolbar with numerous tools. It lets you draw new features on a map, edit existing features, and also delete features.

The tools that the editor dijit provides are as follows:

- Feature selection tools, either to add new selections, add to them further, or to remove from existing selections
- Feature drawing tools, including a tool to delete features from the map
- Tools that let you cut, merge, and clip parts of polylines and polygons
- Undo and redo operations (requires versioned map services)

Snapping manager

Here's a common request you might receive when creating an editing tool: "I'd like a tool that lets me edit this feature based on the lines of this other feature". You could try to code your own tool to select a feature and go through each phase of the feature. Or, with a few additional settings, you could implement the map's **snapping manager**.

The snapping manager imitates ArcMap snapping controls in the browser . As your mouse pointer approaches the corner or edge of a graphic feature, perhaps in a `GraphicsLayer` or a `FeatureLayer`, a new pointer moves over the point on the feature. This shows where you would add a point if you clicked on the map. You can click along a set of points, line vertices, or polygon corners to draw something that lines up perfectly with existing features with this tool.

When loading the snapping manager, there are a few important options that need to be set. Every snapping manager requires a map to snap to. It also requires a graphics layer or a feature layer to load, along with information about its snapping behavior. It should know whether to snap to the edge or vertex of a line or polygon, as well as whether to snap to points of a point feature class. All this information is added in a `layerInfo` array in its constructor options, or can be added later by using the `setLayerInfos()` method.

There are other optional configurable items in the snapping manager. You can tell the snapping manager to always snap to a graphic, or whether you want to control snapping by holding down a key on the keyboard while clicking. You can also configure which keyboard key is the `snapKey`, by loading that property with the appropriate `dojo/keys` constant. Finally, the `tolerance` of a snapping manager refers to the maximum number of pixels the pointer should be from the feature before it snaps to it.

You can see an example of a snapping manager loaded in a JavaScript API in the following code:

```
require([…, "esri/SnappingManager", "dojo/keys", …],
function (…, SnappingManager, dojoKeys …) {

  var propertyLayer = new FeatureLayer({…});
  var sm = new SnappingManager({
    alwaysSnap: false, // default: false
    map: map,
    snapKey: dojoKeys.CTRL, // default: dojoKeys.copyKey
    tolerance: 10, // default: 15
    layerInfo: [{
```

```
      layer: propertyLayer, // this is a featureLayer,
      snapToEdge: false, // default: true
      snapToVertex: true //default: true
    }]
  });
  ...
});
```

The preceding example shows a snapping manager that turns on when the user holds down the *Ctrl* key on a PC (the *Command* key on a Mac). It only snaps to the corners of a line or polygon in the `propertyLayer` feature layer. The `tolerance` for snapping was set to 10 pixels.

Securing editing services

If you're going to open up your data to be edited by the public, you need to be prepared for trouble. From bad data input to malicious attacks, you, as a developer, need to account for things going wrong. Luckily, ArcGIS Server and the ArcGIS API for JavaScript can help.

Restricting user input

I remember a project where we had to let users search for addresses based on a list provided by another system. The other system had no restrictions on what the user could enter. As a result, the address list was anything but normal. On a given street, there could be fifteen different ways the street name could be listed. Some were all caps, while others had "Rd" instead of "Road" Others were misspelled, one m instead of two, and some had too many spaces between the street name and the suffix. Needless to say, the data was poorly constructed and unstandardized.

ArcGIS Server provides some tools to help you restrict user input. Implementing coded value domains and ranges in the geodatabase can help reduce bad input. The attribute inspector honors field properties such as length and data type. You can set default values to limit extra user input in the feature service feature templates,

You can also tie in validation and other controls to make sure the user does not accidently do something like add a phone number to a date column. Dojo comes with user controls such as validation textboxes that limit bad input.

Password protected services

ArcGIS Server also offers a better option when it comes to securing your editing services. If you want to restrict access to editing data, you can demand token-based authentication for map services. A **token** is an encrypted string that contains a user name, an expiration date, and extra information for verification purposes. You need to request a token from `http://myserver/arcgis/tokens`, where myServer refers to your ArcGIS Server web endpoint or web adaptor. You submit the necessary user name and password before having the token added as a cookie on your browser. Tokens are only good for a limited time, which can be adjusted through configurable settings in ArcGIS Server.

These token-based security measures work with both map services and editable feature services. Without the token, you are not able to see the protected map services in the browser. With it, you can explore secured services, query them, and even edit data in them.

Identity manager

The **identity manager** (`esri/IdentityManager`) is used to handle logins and the security of ArcGIS Server and ArcGIS Online services. The identity manager displays a username and password prompt when you attempt to load token-protected services in the browser. Its user interface uses Dojo UI controls, so loading the appropriate Dojo style sheet is necessary to make the identity manager work properly.

Now that we've reviewed some of the editing capabilities ArcGIS Server offers, let's apply what we've learned to an application.

A new mapping application

So, back to our story and our mapping application. We're still waiting for word from the Y2K society about the Census map, but we have a new application we've been asked to work on. It seems that the city of Hollister, California has asked us to put an app together for them. Let's find out what they want.

The city of Hollister wants to create an application that lets citizens report issues in the city. They want citizens to report things like graffiti, sidewalk, curb, and street issues, damaged property, sewer issues, and tree problems, on a map, and also supply additional information. If possible, they want photos of the problems so crews know what to look for.

The file setup

We don't need to create any custom packages because we're going to use the out-of-the-box ArcGIS JavaScript API editing tools,. Instead, we'll create a simple file setup with a `css` and a `js` folder, We'll add our custom `style.css` style sheet in the `css` folder, We'll add our `app.js` file in the `js` folder. We'll also add a folder named `proxy` to handle our proxy service. The file structure should look something like the following:

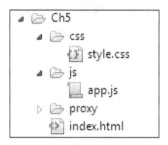

The front page

Let's start with the HTML document. We'll use our basic cookie-cutter site. This time, we'll add Dojo's `claro.css` style sheet. We don't need any custom packages, so we can leave those out of the `dojoConfig` file. We'd like a long column going down the side of the page, and a header part where we'll put the title for the page layout. We'll load the `BorderContainer` with a sidebar design with the taller side columns. We'll add three `ContentPanes` for the header, a leading column for the buttons, and a center region for the map.

```
<!DOCTYPE html>
<html>
    <head>
    <meta http-equiv="Content-Type" content="text/html;
    charset=utf-8" />
    <meta http-equiv="X-UA-Compatible" content="IE=Edge" />
    <meta name="viewport" content="initial-scale=1,
    maximum-scale=1,user-scalable=no" />
    <title>Incident Reporting App</title>
    <meta name="description" content="">
    <meta name="author" content="Ken Doman">
    <link rel="stylesheet" href="http://js.arcgis.com/3.13/
    dijit/themes/claro/claro.css">
```

```
    <link rel="stylesheet" href="https://js.arcgis.com/3.13/
    esri/css/esri.css" />
    <link rel="stylesheet" href="./css/style.css" />
    <script type="text/javascript">
      dojoConfig = {
        async: true,
        isDebug: true
      };
    </script>
    <script src="https://js.arcgis.com/3.13/"></script>
  </head>
  <body class="claro">
    <div id="mainwindow"
      data-dojo-type="dijit/layout/BorderContainer"
      data-dojo-props="design:'sidebar', gutter:false"
      style="width: 100%; height: 100%; margin: 0;">
        <div data-dojo-type="dijit/layout/ContentPane"
        data-dojo-props="region:'top'">
          <h1>Incident Reporting App</h1>
        </div>
        <div id="map" data-dojo-type="dijit/layout/ContentPane"
        data-dojo-props="region:'center'"></div>
        <div id="editpane" style="width: 130px"
        data-dojo-type="dijit/layout/ContentPane"
        data-dojo-props="region:'leading'">
          <div id="editordiv"></div>
        </div>
      </div>
    <script type="text/javascript" src="js/app.js"></script>
  </body>
</html>
```

We'll add some basic styling for the HTML and the body on the `style.css` page. Let's add the following style:

```
html, body {
  width: 100%;
  height: 100%;
  border: 0;
  margin: 0;
  padding: 0;
  box-sizing: border-box;
  font-family: Helvetica, Arial, sans-serif;
}

*, *:before, *:after { box-sizing: inherit;}
```

We've set the `width` and `height` of the HTML and `body` to `100%`, with no border, margin, or padding. We've also changed the font to a common `sans-serif` font, such as `Helvetica`, `Arial`, or just plain `sans-serif`. Finally, we set the elements on the page to be sized using border-box `box-sizing` which makes it easier to work with sizing boxes on the page.

Loading the map

We'll begin writing the code for our application with the page setup. We have a focus area, which is the city of Hollister. For the sake of an easy life, we'll add the city boundaries as an extent:

```
require([
  "dojo/parser", "esri/map", "esri/graphic",
  "esri/geometry/Extent", "esri/dijit/editing/Editor",
  "esri/dijit/editing/TemplatePicker", "esri/tasks/query",
  "dijit/layout/BorderContainer", "dijit/layout/ContentPane",
  "dojo/domReady!"
], function (parser, Map, Graphic, Extent,Editor, TemplatePicker,
  Query) {

  var maxExtent = new Extent({
    "xmin":-13519092.335425414,
    "ymin":4413224.664902497,
    "xmax":-13507741.43672508,
    "ymax":4421766.502813354,
    "spatialReference":{"wkid":102100}
  }),
    map, selected, updateFeature, attInspector;

  parser.parse();

  map = new Map("map", {
    basemap: "osm",
    extent: maxExtent
  });

});
```

In the preceding code, we've loaded the necessary modules and used the `dojo/parser` to parse them. We've added a map with the OpenStreetMap based basemap, and we've created a `maxExtent` to simulate the city boundaries.

Adding the map layers

Now that we have our map, we need to add layers to the map. For the sake of this exercise, we're going to use the San Francisco 311 Feature Service provided by ESRI. We're going to load the feature layer in selection mode, so we only affect the features we click on. We're also going to add the complementary dynamic map service, because we can't see the features without it. We will also set the feature layer selection symbol using a simple marker symbol to color the features we click on:

```
require([
  ...,
  "esri/layers/FeatureLayer",
  "esri/layers/ArcGISDynamicMapServiceLayer",
  "esri/symbols/SimpleMarkerSymbol",
  ...
], function (...,
  FeatureLayer, ArcGISDynamicMapServiceLayer,
  MarkerSymbol, ...
) {
  var maxExtent = ...,
    map, incidentLayer, visibleIncidentLayer;
  ...
  incidentLayer = new
  FeatureLayer("http://sampleserver3.arcgisonline.com/
  ArcGIS/rest/services/SanFrancisco/311Incidents/FeatureServer/0",
  {
    mode: FeatureLayer.MODE_SELECTION,
    outFields:
    ["req_type","req_date","req_time","address","district",
    "status"],
    id: "incidentLayer"
  });

  visibleIncidentLayer = new
  ArcGISDynamicMapServiceLayer(
  "http://sampleserver3.arcgisonline.com/ArcGIS/rest/services/
  SanFrancisco/311Incidents/MapServer");
  ...
  map.addLayers([visibleIncidentLayer, incidentLayer]);
```

When the map layers are added, we can finally interact with them, both as a user and as a developer. We'll add an event listener called `startEditing()` to the map's `layers-add-result` event. We'll set up the editing events for the feature layer there. We'll add a map click event that draws a feature if something has been selected from the menu on the side of the page. Be sure to add this after the layers are defined, but before they are added to the map.

```
var …, incidentLayer, visibleIncidentLayer, selected;

visibleIncidentLayer = …;

function startEditing () {
  var incidentLayer = map.getLayer("incidentLayer");
  // add map click event to create the new editable feature
  map.on("click", function(evt) {
    // if a feature template has been selected.
    if (selected) {
      var currentDate = new Date();
      var incidentAttributes = {
        req_type: selected.template.name,
        req_date:(currentDate.getMonth() + 1) + "/" +
        currenDate.getDate() + "/" + currentDate.getFullYear(),
        req_time: currentDate.toLocaleTimeString(),
        address: "",
        district: "",
        status: 1
      };
      var incidentGraphic = new Graphic(evt.mapPoint,
      selected.symbol, incidentAttributes);
      incidentLayer.applyEdits([incidentGraphic],null,null)
    }
  });

  incidentLayer.setSelectionSymbol(
    new MarkerSymbol({color:[255,0,0]})
  );

  map.infoWindow.on("hide", function() {
    incidentLayer.clearSelection();
  });
}
```

```
incidentLayer.on("edits-complete", function() {
  visibleIncidentLayer.refresh();
});

map.on("layers-add-result", startEditing);
map.addLayers([visibleIncidentLayer, incidentLayer]);
```

In the preceding code, we've created a `callback` function called `startEditing()`. This causes the application to add a new graphic to the editable feature layer whenever the map is clicked. Default attributes and a symbol are applied to the editable feature. The editable feature layer clears its selection whenever the popup is hidden. Also, when the edits are complete, the visible layer is refreshed with the new data. The `startEditing()` method is assigned to run when a group of layers are added, which causes the layers to be added to the map.

Using the proxy

If you try to load the map right now, you may get an error. If you don't get it now, you might get it when you try to save changes on the map. The reason is that these editing operations often require a proxy application to handle data which is too large to fit in the approximately 2,048 character limit of most browser get requests.

 You can follow the instructions ESRI provides to set up a proxy service at `https://developers.arcgis.com/javascript/jshelp/ags_proxy.html`.

Proxies come in three varieties, based on your application environment. ESRI provides proxy services in PHP, Java, and .Net. We'll add a reference to the proxy in our application. This example shows how it's done with a .Net based proxy:

```
require([
  ...,
  "esri/config",
  "esri/layers/FeatureLayer",
  "esri/layers/ArcGISDynamicMapServiceLayer",
  "esri/symbols/SimpleMarkerSymbol",
  ...
], function (..., esriConfig, ...) {
  ...
  // set up proxy for the featureLayer
  esriConfig.defaults.io.proxyUrl = "./proxy/proxy.ashx";

  incidentLayer = ...
```

Finding the user's location

Our client requested that the app provide the user with the ability to find them on the map, should they be using a mobile device or a laptop on Wi-Fi. We can provide that functionality by adding an ArcGIS dijit called `LocateButton`. We load the module in our application, initialize it when the map is ready, and it's good to go. The code to load it should look something like this:

```
require([…, "esri/dijit/LocateButton", …
], function (…, LocateButton, …) {
  …
  function startEditing() {
    // add the Locate button
    var locator = new LocateButton({map: map}, "locatebutton");
  }
  …
});
```

If we insert a `<div>` with an `id` of `locatebutton` inside the map `ContentPane`, and view the page in our browser, we'll see the locate button above the map, pushing the map down. We'd much rather locate it near the other zoom in and out buttons. We'll add the following styling to the `style.css` sheet to achieve that:

```
.LocateButton {
  position: absolute;
  left: 29px;
  top: 120px;
  z-index: 500;
}
```

The template picker

For our application, we're going to use the ArcGIS JavaScript API's template picker to select incident point types to add to the map. We'll load them in the side pane on the page, and make them one column wide to add features. We'll pass that feature template to the `selected` variable when the feature template is selected. Finally, we'll load all this when both the map and the feature layers have loaded:

```
function generateTemplatePicker(layer) {
  console.log("layer", layer);
  var widget = new TemplatePicker({
    featureLayers: [ layer ],
    rows: layer.types.length,
    columns: 1,
```

```
          grouping: false,
          style: "width:98%;"
      }, "editordiv");

      widget.startup();

      widget.on("selection-change", function() {
          selected = widget.getSelected();
          console.log("selected", selected);
      });
  }
...
function startEditing () {
    var incidentLayer = map.getLayer("incidentLayer");
    generateTemplatePicker(incidentLayer);
    ...
```

The attribute inspector

Now that we are able to add new features to the map, we need a way to edit the
content of those features. To do that, we'll add the attribute inspector. We're going
to initialize the attribute inspector and tie it to the map's `infoWindow`.

```
require([…,
    "dojo/dom-construct",
    "esri/dijit/AttributeInspector",
    "dijit/form/Button",
    …
], function (…, domConstruct, AttributeInspector, Button, …) {
    var maxExtent = …,
        map, incidentLayer, visibleIncidentLayer, selected,
        attInspector, updateFeature;
...
function generateAttributeInspector(layer) {
    var layerInfos = [{
        featureLayer: layer,
        showAttachments: true,
        isEditable: true,
    }];

    attInspector = new AttributeInspector({
        layerInfos: layerInfos
    }, domConstruct.create("div", null, document.body));
```

```
    attInspector.startup();

    //add a save button next to the delete button
    var saveButton = new Button({ label: "Save", "class":
    "saveButton"});
    domConstruct.place(saveButton.domNode,
    attInspector.deleteBtn.domNode, "after");

    saveButton.on("click", function(){
      updateFeature.getLayer().applyEdits(
        null, [updateFeature], null
      );
    });

    attInspector.on("attribute-change", function(evt) {
      //store the updates to apply when the save button is clicked
      updateFeature.attributes[evt.fieldName] = evt.fieldValue;
    });

    attInspector.on("next", function(evt) {
      updateFeature = evt.feature;
      console.log("Next " + updateFeature.attributes.objectid);
    });

    attInspector.on("delete", function(evt){
      evt.feature.getLayer().applyEdits(
        null, null, [updateFeature]
      );
      map.infoWindow.hide();
    });

    if (attInspector.domNode) {
      map.infoWindow.setContent(attInspector.domNode);
      map.infoWindow.resize(350, 240);
    }

}
...
function startEditing () {
  var incidentLayer = map.getLayer("incidentLayer");
  generateTemplatePicker(incidentLayer);
  generateAttributeInspector(incidentLayer);
...
```

We'll need to add a little positioning style to the save button on the attribute inspector. We'll add this entry to position the save button in the `style.css` sheet so that it doesn't overlap the delete button.

```css
.saveButton {
  margin: 0 0 0 48px;
}
```

Now that the attribute inspector is loaded, we can incorporate it into the click events for both the map layer and the incident layer. We'll create a `showInspector()` function that accepts a map click event. It will query the `incidentLayer` for any features in that location, and pull up a map `infoWindow` with the attribute inspector inside. It will also assign the selected graphic (if any) to:

```javascript
…
    function showInspector(evt) {
       var selectQuery = new Query(),
         point = evt.mapPoint,
         mapScale = map.getScale();

       selectQuery.geometry = evt.mapPoint;

       incidentLayer.selectFeatures(selectQuery,
       FeatureLayer.SELECTION_NEW, function (features) {
         if (!features.length) {
           map.infoWindow.hide();
           return;
         }

         updateFeature = features[0];

         map.infoWindow.setTitle(updateFeature.getLayer().name);
         map.infoWindow.show(evt.screenPoint,
         map.getInfoWindowAnchor(evt.screenPoint));
       });
    }
…
    function startEditing() {
       …
       map.on("click", function (evt) {
         …
         if (selected) {
           …
           incidentLayer.applyEdits([incidentGraphic],null,null)
             .then(function () {
```

```
        showInspector(evt);
    });

  } else {
    showInspector(evt);
  }
  ...
  });
...
    incidentLayer.on("click", showInspector);
```

The result of the preceding code is shown in the following screenshot:

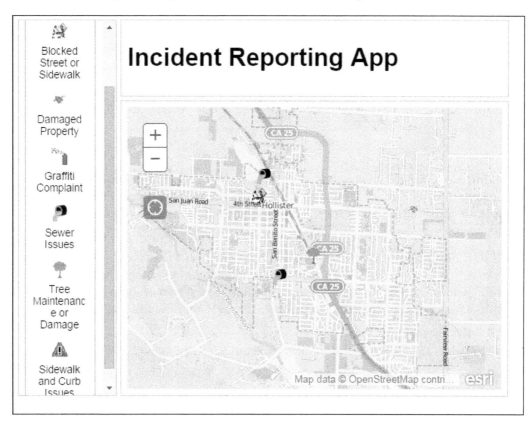

Securing the application

Now that we have a working incident reporting application, it's time to think about how we can secure the application. It's a public application, so the public should be allowed to submit problems. However, we don't want data that doesn't fit our data schema, or our representative boundaries.

One way we can secure our application from bad input is to restrict the locations where we accept changes. We don't want employee time wasted investigating complaints logged outside the city, state, or even country. We can do this by using the city extent supplied at the beginning of the application. We can test if the clicked point is inside the city extent in the click event, and notify the client if it's not. That should look something like the following:

```
...
map.on("click", function (evt) {
  // if the clicked point isn't inside the maxExtent
  if (!maxExtent.contains(evt.mapPoint)) {
    alert("Sorry, that point is outside our area of interest.");
    return; // disregard the click
  }
...
```

Speaking of working with extents, we could also lock the selection buttons when navigating outside the city area. This would alert the user that we aren't accepting complaints outside the city proper. Of course, we should notify the user why they are locked out.

We'll start by adding blocking content and notifications in the HTML. We'll add two divs to the page, a `<div>` with the id outsidemessage in the map, and a div with the id `blockerdiv` next to the editor div. We'll leave the two of them hidden by default, by adding the inline style `display: none`. It should look like the following:

```
...
<div id="map" data-dojo-type="dijit/layout/ContentPane"
  data-dojo-props="region:'center'" >
  <div id="locatebutton"></div>
  <div id="outsidemessage" style="display:none;">
    <p>Sorry, but you have navigated outside our city. Click on this
    message to get back to the city.</p>
  </div>
</div>
<div id="editpane" style="width: 130px"
  data-dojo-type="dijit/layout/ContentPane"
```

```
    data-dojo-props="region:'leading'">
    <div id="editordiv"></div>
    <div id="blockerdiv" style="display:none;"></div>
</div>
```

We'll add the following styling to the `style.css` file to style these items. The outside message will be gray block floating in the lower middle portion of the map, with text big enough to read, and with rounded corners (because lots of people like rounded corners). The blocking `div` will be positioned directly on top of the template picker buttons. The blocking `div` will be light gray, semi-transparent, and cover the entire template picker when visible.

```css
#blockerdiv {
  width: 100%;
  background: rgba(188, 188, 188, 0.6);
  position: absolute;
  top: 0;
  left: 0;
  bottom: 0;
  z-index: 500;
}

#outsidemessage {
  position: absolute;
  bottom: 40px;
  left: 50%;
  width: 200px;
  height: auto;
  margin: 0 0 0 -100px;
  background: rgba(255,255,255,0.8);
  padding: 8px;
  z-index: 500;
  border-radius: 8px;
  text-align: center;
  font-weight: bold;
}
```

We'll add some code to our `app.js` file to handle the visibility of these two nodes. We can listen for changes in the map's extent. When the map's extent is outside the city extent, and they no longer intersect, both the message div and the blocker div will be made visible (`display: block;`). If the user can see some of the extent of the viewing area, the `div` objects will be hidden again (`display: none;`). It should look like this:

```
require([…, "dojo/dom-style", …
], function (…, domStyle, …) {

function onMapExtentChange (response) {
  if (!response.extent.intersects(maxExtent)) {
    // show these blocks if the map extent is outside the
    // city extent
    domStyle.set("blockerdiv", "display", "block");
    domStyle.set("outsidemessage", "display", "block");
  } else {
    // hide these blocks if the max Extent is visible within
    // the view.
    domStyle.set("blockerdiv", "display", "none");
    domStyle.set("outsidemessage", "display", "none");
  }
}

map.on("extent-change", onMapExtentChange);
…
```

We'll also add an event handler to the outside message `div` that lets the user click to go back to the starting location for the map. We'll load the `dojo/on` event to handle the `click` event:

```
require([…, "dojo/dom", "dojo/on", …
], function (…, dojoDom, dojoOn, …) {
  …
  dojoOn(dojoDom.byId("outsidemessage"), "click", function () {
    map.setExtent(maxExtent);
  })
  …
});
```

Now, when we load our application and pan our way outside the city limits, the following message should appear:

Limiting data entry

As well as stopping bad user input from clicks, we should also consider stopping bad user input from text editing. We can pull that off by modifying the `layerInfos` array assigned to the attribute inspector. We'll start by getting rid of the delete button, since we don't want citizens deleting everybody else's complaints. We'll also modify the `fieldInfos` list and set some of the fields to display only when they are edited in the attribute inspector. In this case, we'll leave the `req_type`, address, and district tabs open for editing.

```
var layerInfos = [{
    featureLayer: layer,
    showAttachments: true,
    isEditable: true,
    showDeleteButton: false,
    fieldInfos: [
      {'fieldName': 'req_type', 'isEditable':true, 'tooltip':
      'What\'s wrong?', 'label':'Status:'},
      {'fieldName': 'req_date', 'isEditable':false, 'tooltip':
      'Date incident was reported.', 'label':'Date:'},
```

```
          {'fieldName': 'req_time',
          'isEditable':false,'label':'Time:'},
          {'fieldName': 'address', 'isEditable':true,
          'label':'Address:'},
          {'fieldName': 'district', 'isEditable':true,
          'label':'District:'},
          {'fieldName': 'status', 'isEditable':false,
          'label':'Status:'}
        ]
    }];
```

These are a few of the simple things we can do to help secure our application against unwanted results, and yet still make the application user-friendly to the general public.

Summary

In this chapter, we've examined the tools and processes that ArcGIS Server and the ArcGIS API for JavaScript provide to make web editing possible. We looked at what goes into an editable feature service. We also looked into the various widgets that come with the ArcGIS JavaScript API for adding new features, editing geometries, and editing property attributes. We finished by creating an application that uses the editing tools to create an incident reporting application, with which users can report problems in the city on the map.

In the next chapter, we'll take the existing data and add a graphical twist.

6
Charting Your Progress

Displaying data geographically provides users with locational awareness, but some want to see more than just dots on a map and some numbers. They want to see how the data in each location compares, both across the map, and within a location. Other methods of displaying data, such as charts and graphs, can provide additional information.

Charts and graphs are big business. Companies spend millions of dollars creating executive dashboards, which are a mix of charts and graphs connected to company data and metrics. They work because humans aren't as good at processing large, abstract numbers as computers, but they do better at processing data visually. Good charts and graphs provide comparable data at a glance, in a way anyone can understand.

In this chapter, we shall learn:

- How to create charts and graphs using tools provided by the ArcGIS JavaScript API and the Dojo framework
- How to implement the same charts and graphs using D3.js
- How to add an external library such as D3.js as an AMD module

Mixing graphs into our maps

As we have learned before, the **ArcGIS JavaScript API** contains more than tools to create maps and text. Built on top of the Dojo framework, the ArcGIS API comes with many user controls and widgets to help you present data. We can create dynamic tables, charts, graphs, and other data visualizations.

But you are not limited to the charting and graph tools provided by API. Using Dojo's AMD style, you can incorporate other libraries outside the framework into your widget's build, and load them as they are needed. If you work with team members who are more familiar with a library like D3.js, you can load the library asynchronously into your widget and let the other person develop the graphics.

In this chapter, we'll explore both internal and external graphics libraries to add graphs to our data. We'll use the dojox/charting (http://dojotoolkit.org/reference-guide/1.10/dojox/charting.html) modules packaged with the ArcGIS JavaScript API, We'll also implement the graphs with D3.js (http://d3js.org), a popular data visualization library.

Our story continues

Our clients from the Y2K society called with another request. They don't like the tables we've added to our census popups. All the big numbers overwhelm some of the users. They would rather see the data represented in graphs, so that the user can see the relationships within the data.

The Y2K society specifically requested pie charts for ethnic and gender data. We can use any color for the ethnic data, but for gender, they made specific color requests. They want to use a horizontal bar chart for the age data because they've seen population data shown that way and they liked the look. They would like some of the other data graphed and charted as well, but they are willing to leave how that's done to our discretion.

For this round, we're going to try two different approaches and see which one the client prefers. We'll create charts using the two libraries with the same data, add them to the census data popups, and see which one the client prefers.

Using dojox charting

We should first look inside the ArcGIS API for JavaScript to see what it has to offer. We can access a host of charting resources with dojox/charting. The modules allow you to draw line graphs, pie charts, bar charts, and a whole host of other graphs and charts in a browser. It contains numerous canned themes to show your data, and it can be extended with your custom themes as well. The charting libraries can render in **Scalable Vector Graphics (SVG)**, **Vector Markup Language (VML)**, Silverlight, and Canvas, making them both progressive and backwards-compatible for older browsers like IE7.

Like most Dojo components, `dojox/charting` can render charts either declaratively within your HTML, or programmatically through JavaScript. Declarative charting takes advantage of the `data-dojo` attributes. In the exercises that follow, we'll explore the programmatic examples, since they are more dynamic and easier to troubleshoot when things go wrong.

Creating a chart in JavaScript

There is no one module for `dojox/charting` that will take care of all your graphing needs. These four major classes of modules within `dojox/charting` can be loaded to create a unique chart or graph, and add new looks and functionality:

- The chart object
- The chart style
- The chart theme
- The chart actions and effects

The following is a sample loading of each of the four module types, in order:

```
require([
"dojox/charting/Chart",   // object
"dojox/charting/plot2d/Pie",    // style
"dojox/charting/themes/MiamiNice",    // theme
"dojox/charting/action2d/Highlight",   // action
"dojox/charting/action2d/Tooltip",   // action
"dojo/ready"],
  function(Chart, Pie, MiamiNice, Highlight, Tooltip,
  ready){
  ready(function(){
    var myChart = new Chart("myChart ");
    myChart.setTheme(MiamiNice)
     .addPlot("default", {
        type: Pie,
        font: "normal normal 11pt Tahoma",
        fontColor: "black",
        labelOffset: -30,
        radius: 80
    }).addSeries("Series A", [
      {y: 3, text: "Red", stroke: "black", tooltip: "Red
      Alert"},
      {y: 5, text: "Green", stroke: "black", tooltip:
      "Green Day"},
```

```
        {y: 8, text: "Blue",  stroke: "black", tooltip: "I'm
        Blue!"},
        {y: 13, text: "Other", stroke: "black", tooltip: "A
        bit different"}
      ]);
      var anim_a = new Highlight(myChart, "default");
      var anim_b = new Tooltip(myChart, "default");
      myChart.render();
    });
  });
```

The chart object

The chart object, loaded with the `dojox/charting/Chart` module, is the main object you'll use to create and modify your chart. Almost all of your your customizations will be run through this object. The chart object is created with a reference to an HTML element on the page, either by a node or by a string matching the id of a node. In the following code, you can see an example that shows the creation of a simple chart:

```
require(["dojox/charting/Chart", "dojox/charting/axis2d/Default",
  "dojox/charting/plot2d/Lines", "dojo/ready"],
  function(Chart, Default, Lines, ready){
  ready(function(){
    var chart1 = new Chart("fibonacci");
    chart1.addPlot("default", {type: Lines});
    chart1.addAxis("x");
    chart1.addAxis("y", {vertical: true});
    chart1.addSeries("Series 1", [1, 1, 2, 3, 5, 8, 13, 21]);
    chart1.render();
  });
});
```

In the preceding code, a line graph is produced with values in a series that increase from 1 to 21.

The construction of the chart object can also accept an options object. In these options, you can add the map title, and control elements, such as the title text, position, font, color, and the gap between the title and the graph.

The `dojox/charting` library also includes a 3D charting library, called `dojox/charting/Chart3D`. The object can render three dimensional charts and graphs, which can be rotated and panned to get a good perspective on your data. In the following code, you can see an example of a 3D plotted bar chart:

```
require(["dojox/charting/Chart3D", "dojox/charting/plot3d/Bars",
  "dojox/gfx3d/matrix", "dojo/ready"
], function(Chart3D, Bars, m, ready){
  ready(function(){
    var chart3d = new Chart3D("chart3d", {
        lights: [{direction: {x: 6, y: 6, z: -6}, color: "white"}],
        ambient:  {color:"white", intensity: 3},
        specular: "white"
      },
    [m.cameraRotateXg(10), m.cameraRotateYg(-10), m.scale(0.8),
    m.cameraTranslate(-50, -50, 0)]),
      bars3d_a = new Bars(500, 500, {gap: 8, material: "red"}),
      bars3d_b = new Bars(500, 500, {gap: 8, material: "#0F0"}),
      bars3d_c = new Bars(500, 500, {gap: 8, material: "blue"});
    bars3d_a.setData([3, 5, 2, 4, 6, 3, 2, 1]);
    chart3d.addPlot(bars3d_a);

    bars3d_b.setData([5, 6, 4, 2, 3, 1, 5, 4]);
    chart3d.addPlot(bars3d_b);

    bars3d_c.setData([4, 2, 5, 1, 2, 4, 6, 3]);
    chart3d.addPlot(bars3d_c);

    chart3d.generate().render();
  });
});
```

In the preceding code, a 3D bar graph has been produced with three sets of data colored red, green, and blue. These values are then viewed through a camera that is rotated somewhat to add perspective to the images.

The chart style

The chart style describes what kind of chart we're creating. It defines whether we're loading the data as a line chart or a bar chart, a pie chart or a scatter plot. For two dimensional charts, you'll find these styles in the `dojox/charting/plot2d` folder. Chart styles can be grouped into five main categories, as follows:

- **Lines**: These are the typical line charts that may or may not show the individual data points.

- **Stacked Lines**: Similar to line charts, but the heights are stacked on top of each other. These allow the user to compare the combined effect of plotted data over time, as well as changes in ratios.

- **Bars**: Compare values by the width of rows on a graph.

- **Columns**: Compare quantities by their related column heights.

- **Miscellaneous**: When other charts cannot be grouped together in a category like the previous ones, they fall into this category. This group includes pie charts, scatter plots, and bubble plots.

If you are using the 3D charts, the styles for these charts can be found in the `dojox/charting/plot3d` folder. To take full advantage of the 3D styles, it is best to load the `dojox/gfx3d/matrix` module for 3D graphic effects. The `matrix` module allows you to rotate the 3D graph in order to get a good perspective of the 3D charts.

The chart theme

Chart themes describe the colors, shading, and text formatting of text elements within your charts and graphs. The Dojo framework comes with a large number of predefined themes that you can choose from, in `dojox/charting/themes`.

 You can see what the different themes look like by going to `http://archive.dojotoolkit.org/nightly/checkout/dojox/charting/tests/test_themes.html`.

The following example is code which loads a chart with a `MiamiNice` theme. In this example, we have loaded a line chart with an x and y axis. We set the theme to `MiamiNice` using the `setTheme()` method. Then, we added the series of numbers to plot and render the chart:

```
require(["dojox/charting/Chart", "dojox/charting/axis2d/Default",
    "dojox/charting/plot2d/Lines", "dojox/charting/themes/MiamiNice",
    "dojo/ready"],
    function(Chart, Default, Lines, MiamiNice, ready){
    ready(function(){
```

```
        var chart1 = new Chart("simplechart");
        chart1.addPlot("default", {type: Lines});
        chart1.addAxis("x");
        chart1.addAxis("y", {vertical: true});
        chart1.setTheme(MiamiNice);
        chart1.addSeries("Fibonacci", [1, 1, 2, 3, 5, 8, 13, 21]);
        chart1.render();
    });
});
```

If you don't find a theme that works for you, or if you have specific colors and styling that you need to adhere to in your application design, you can use the SimpleTheme object to help define your custom theme. SimpleTheme is based on the GreySkies theme, but can be extended with other colors and any formatting you choose. You do not need to define every attribute of your theme, since SimpleTheme applies whatever defaults you haven't overridden with your custom style. You can see a sample of the code that implements SimpleTheme here:

```
var BaseballBlues = new SimpleTheme({
    colors: [ "#0040C0", "#4080e0", "#c0e0f0", "#4060a0", "#c0c0e0"]
});
myChart.setTheme(BaseballBlues);
```

Themes typically use no more than five colors in their palette. If you need to add more colors for a set of data, push() a color hex string into the theme's .color array, but do it prior to setting the theme of the chart.

Chart actions and effects

Creating appealing charts and graphs might be fun for you, but users in the modern web era expect to interact with the data. They expect chart elements to glow, grow, and change color when they hover over them. They expect things to happen when they click on a pie chart.

The dojox/charting/action2d contains chart actions and effects that make charts more educational and interactive. You don't have to overdo the actions and make your graph do everything. You can simply apply the events you need to get the effect across to the users. The following is a list of the basic actions and effects, along with descriptions:

- Highlight: This adds a highlight to the chart or graph element you select.

- Magnify: This lets you magnify a portion of the chart or graph for easier viewing.

- `MouseIndicator`: You can drag your mouse over features on the graph to show more data.
- `MouseZoomAndPan`: This lets you zoom and pan over your graph using the mouse. The scroll wheel zooms in and out, while click and drag lets you pan around the graph.
- `MoveSlice`: When using a pie chart, clicking on a slice can move it out from the rest of the chart.
- `Shake`: This creates a shaking action on an element on the chart
- `Tooltip`: Hovering the mouse cursor over a chart element shows more information.
- `TouchIndicator`: This provides touch actions that display data on charts.
- `TouchZoomAndPan`: This gives you zoom and pan ability using touch gestures.

Unlike chart styles and themes, where you attach the chart component to the chart object, chart actions are called separately. The chart action constructor loads the new chart as the first argument, and optional parameters for the second argument. Note that the actions are created before the chart is rendered. You can see an example in the following code:

```
require(["dojox/charting/Chart",
  ...,
  "dojox/charting/action2d/Tooltip"],
  function(Chart, ..., Tooltip){
    var chart = new Chart("test");
    ...
    new Tooltip(chart, "default", {
       text: function(o){
          return "Population: "+o.y;
       }
    });
    chart.render();
});
```

In the preceding example, a chart was created, and a tooltip was added showing the population as you hovered over a graph feature.

Using Dojox Charts in popups

Combining the dojox/charting modules with the ArcGIS API for JavaScript provides many ways to display data. One way to deliver feature data through charts is through the map's `infoWindow`. The infoWindow uses an HTML template for its content, and that can provide the hooks we need to attach our graphs.

One issue when adding graphs to the infoWindow is determining when to draw the graph. Thankfully, there's an event for that. The map's `infoWindow` fires a `selection-changed` event whenever the selected feature graphic is changed, either by clicking on another graphic, or by clicking on the next and previous buttons. We can assign an event listener to that event, look at the selected graphic and, if it has the data we need, we can draw the graphs.

Using Dojo Charts in our application

Our census application from previous chapters could use some visual appeal when it comes to presenting data. We'll make our first attempt at adding charts and graphs using the `dojox/charting` library. We'll apply the graphs to the map popup whenever the user clicks on a census block group, county, or state. The census blocks don't have enough information for us to graph.

Loading the modules

Since our graphs are currently limited to our census application, we need to update the modules in our custom `y2k/Census` module definition:

1. We'll start by adding `dojo/on` to handle the map popup events.

2. We'll add the default chart object along with a pie chart and a bar chart module.

3. We'll add the `PrimaryColors` theme and `SimpleTheme` to create our own custom color template.

4. Finally, we'll add a highlight and a tooltip action to let the user read the results when they hover over parts of the graphs.

5. It should look a bit like the following:

```
define([…
  "dojo/on",
  "dojox/charting/Chart",
  "dojox/charting/plot2d/Pie",
  "dojox/charting/plot2d/Bars",
  "dojox/charting/action2d/Highlight",
  "dojox/charting/action2d/Tooltip",
  "dojox/charting/themes/PrimaryColors",
  "dojox/charting/SimpleTheme",.
], function (…, dojoOn, Chart, Pie, Bars, Highlight,
  Tooltip, PrimaryColors, SimpleTheme, …) { … });
```

Preparing the popup

As part of our plan, we want the charts and graphs to render in the map's `infowindow` when the features are clicked. We're only interested in showing the charts and graphs for the currently selected feature, so we'll add an event handler to run every time the `infoWindow` object's `selection-change` event fires. We'll call it `_onInfoWindowSelect()`. After we write a stub function for that in our `Census.js` module, we'll add the event handler in the `_onMapLoad()` method. We then know the map and its popup are available. It should look something like the following code:

```
_onMapLoad: function () {
  …
  dojoOn(this.map.infoWindow, "selection-change", lang.hitch(this,
    this._onInfoWindowSelect));
},
  …
_onInfoWindowSelect: function () {
  //content goes here
}
```

The `infoWindow` object's `selection-change` event fires when features are both added and removed from the selection. When we examine the `infoWindow` object's selected feature, we must test to find out if it contains a feature. If one is present, we can process that feature's attributes and add the related graphics to the popup. The `infoWindow` function should look like the following:

```
_onInfoWindowSelect: function () {
  var graphic = this.map.infoWindow.getSelectedFeature(),
    ethnicData, genderData, ageData;
  if (graphic && graphic.attributes) {
    // load and render the ethnic data
    ethnicData = this.ethnicData(graphic.attributes);
    this.ethnicGraph(ethnicData);
    // load and render the gender data
    genderData = this.genderData(graphic.attributes);
    this.genderGraph(genderData);
    // load and render the age data
    ageData = this.ageData(graphic.attributes);
    this.ageGraph(ageData);
  }
},
  …
ethnicData: function (attributes) { },
ethnicGraph: function (data) { },
```

```
genderData: function (attributes) { },
genderGraph: function (data) { },
ageData: function (attributes) { },
ageGraph: function (data) { }
```

Updating the HTML template

In order to add graphs to our popups, we need to update the HTML templates to include element IDs. The JavaScript code then looks for a place to render the graph and we can tell it to render it in the element where the `id` is added. Open `CensusBlockGroup.html` to look at the popup template. Find the *Ethnic groups* section and delete the entire table underneath. You can comment it out for testing purposes, but we don't want everybody to download all that wasted content when we put this application into production. Replace the table with a `div` that has an `id` equal to `ethnicgraph`. It should look like the following:

```
...
<b>Ethnic Groups</b>
<div id="ethnicgraph"></div>
...
```

Repeat the same under the `Males/Females` and the `Ages` sections, replacing those tables with `div` elements identified as `gendergraph` and `agegraph` respectively. If you choose to show other graphs, follow the same guidelines. Repeat with the `CountyCensus.html` and the `StateCensus.html` templates as well.

Processing the data

If you look back through the examples from other `dojox/charting` operations, you'll notice how data is added to the chart in an array. However, the data we get from the map service isn't in that format. We need to process the attribute data into a format that the `dojox/charting` modules can use.

When passing data objects into `dojox/charting` charts and graphs, the graphs expect data to be plottable with x and y properties. Since we're not comparing value changes over time or some other independent variable, we will add the numeric populations to our dependent variable y. The value of the tooltip text can be assigned to the JSON tooltip data property. You can see the resulting function in the following code:

```
...
formatAttributesForGraph: function (attributes, fieldLabels) {
  var data = [], field;
  for (field in fieldLabels) {
    data.push({
      tooltip: fieldLabels[field],
```

```
      y: +attributes[field]
    });
  }
  return data;
},
  ...
```

 The + symbol in front of the attribute in the population objects is a shortcut to convert a value into a number, if it isn't one already. You get the same effect using the `parseInt()` or `parseFloat()` methods.

Now that we're able to transform our data into a format useable for our graph widget, we can call our `ethnicData()`, `genderData()`, and `ageData()` methods. We'll extract the data we need from the feature attributes and put it in an array format to be used by the `chart` module.

Parsing the ethnic data

We're interested in extracting the ethnic makeup of the population in the census area. We're interested in the WHITE, BLACK, AMER_ES, ASIAN, HAWN_PI, HISPANIC, OTHER, and MULT_RACE fields that are present in the state, county, and block group feature classes. Since we have a lot of fields that may or may not be in the feature class, and we'll be adding them the same way, we'll create an array of field names and the corresponding labels we want to add. See the following code:

```
...
ethnicData: function (attributes) {
  var data = [],
      fields = ["WHITE", "BLACK", "AMERI_ES", "ASIAN", "HAWN_PI",
      "HISPANIC", "OTHER", "MULT_RACE"],
      labels = ["Caucasian", "African-American", "Native American
      /<br> Alaskan Native", "Asian", "Hawaiian /<br> Pacific
      Islander", "Hispanic", "Other", "Multiracial"];
},
  ...
```

Now that we have the fields and labels, let's add the information we need to the data array. The `dojox/charting` library expects graphical data in either a numerical list or in a JSON object with a specific format. Since we want to add labels to our data in a pie chart, we'll create the complex objects:

```
...
ethnicData: function (attributes) {
  var fieldLabels = {
    "WHITE": "Caucasian",
    "BLACK": "African-American",
```

```
    "AMERI_ES":"Native American /<br> Alaskan Native",
    "ASIAN": "Asian",
    "HAWN_PI":"Hawaiian /<br> Pacific Islander",
    "HISPANIC": "Hispanic", "OTHER": "Other",
    "MULT_RACE": "Multi-racial"
  }
  return this.formatAttributesForGraph(attributes, fieldLabels);
},
...
```

Parsing the gender data

We'll calculate the gender data in a similar way. We are only interested in the MALES and FEMALES fields in the feature attributes. We're going to add them to the list of JSON objects with the same format as shown in the preceding code. It should look like the following:

```
genderData: function (attributes) {
  var fieldLabels = {
    "MALES": "Males", "FEMALES", "Females"
  }
  return this.formatAttributesForGraph(attributes, fieldLabels);
},
...
```

Parsing the age data

We'll perform the same style of data manipulation for our ageData() method as we did with the ethnicData() method. We'll collect the census data if it's available for ages less than 5, 5-17, 18-21, 22-29, 30-39, 40-49, 50-64, and 65 and up. We'll then add the appropriate tooltip labels and return the resulting formatted data array. It should look as follows:

```
ageData: function (attributes) {
  var fieldLabels = {
    "AGE_UNDER5": "&lt; 5", "AGE_5_17": "5-17", "AGE_18_21": "18-21",
    "AGE_22_29": "22-29", "AGE_30_39": "30-39", "AGE_40_49": "40-49",
    "AGE_50_64": "50-64", "AGE_65_UP": "65+"
  };
  return this.formatAttributesForGraph(attributes, fieldLabels);
},
```

Showing the results

Now that we have the results in a format that we can use for the charts, we can load them into our charts. Our ethnic and gender graphs are both pie graphs, while the age graph is a horizontal bar graph. Let's look at what it takes to construct each. Any extra graphs you want to create with the rest of the data can be done in your own time.

Ethnic graph

We want a pie chart that fits within the popup for the ethnic graph. A radius of 90 pixels should fit nicely within the popup. We're going to set the theme of the graph using `PrimaryColors`, one of the default themes in `dojox/charting`. We'll also add the pie charting ability to the chart, and add the tooltip and highlight animations when the user hovers over the data. Finally, we'll render the ethnic pie chart:

```
ethnicGraph: function (data) {
  var ethnicChart = new Chart("ethnicgraph");
  ethnicChart.setTheme(PrimaryColors)
    .addPlot("default", {
      type: Pie,
      font: "normal normal 11pt Tahoma",
      fontColor: "black",
      radius: 90
  }).addSeries("Series A", data);
  var anim_a = new Tooltip(ethnicChart, "default");
  var anim_b = new Highlight(ethnicChart, "default");
  ethnicChart.render();
},
```

When the application draws the ethnic graph, it should look like the following image:

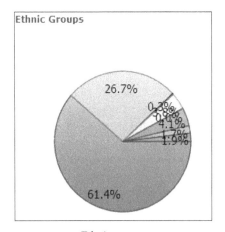

Ethnic groups

Gender graph

For the gender graph, we're going to set up a similar pie graph to the ethnic graph. But, before we do, we'll load a new theme to work with. We'll create a `genderTheme` constructor from the `SimpleTheme` constructor, and add light pink for females, and light blue for males. We'll then create the chart, add the new theme, and add everything else like we did in the ethnic graph. You can see this in the following code:

```
genderGraph: function (data) {
  var genderTheme = new SimpleTheme({
    colors: ["#8888ff", "#ff8888"]
  }),
  genderChart = new Chart("gendergraph");

  genderChart.setTheme(genderTheme)
    .addPlot("default", {
      type: Pie,
      font: "normal normal 11pt Tahoma",
      fontColor: "black",
      radius: 90
  }).addSeries("Series A", data);
  var anim_a = new Tooltip(genderChart, "default");
  var anim_b = new Highlight(genderChart, "default");
  genderChart.render();
},
```

When the application draws the gender graph, it should look something like the following image:

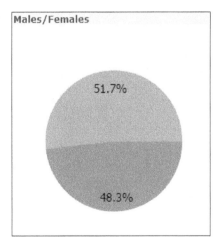

Age graph

We'll create a bar chart graph for the age graph to show the age demographics. Unlike the pie chart, the bar chart doesn't care about the radius, but prefers to know how long the bars can grow (maxBarSize), and how far to set them apart (gap). We'll go ahead and use the PrimaryColors theme again for this object:

```
ageGraph: function (data) {
  var ageChart = new Chart("agegraph");
  ageChart.setTheme(PrimaryColors)
    .addPlot("default", {
      type: Bars,
        font: "normal normal 11pt Tahoma",
        fontColor: "black",
        gap: 2,
        maxBarSize: 220
    }).addSeries("Series A", data);
  var anim_a = new Tooltip(ageChart, "default");
  var anim_b = new Highlight(ageChart, "default");
  ageChart.render();
}
```

When you draw the ageChart, it should look something like the following image:

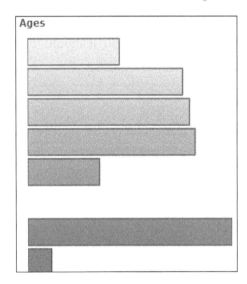

Introducing D3.js

You can branch out beyond the ArcGIS JavaScript API and Dojo if you want to create jaw-dropping graphics. One popular JavaScript library you could use to create charts, graphs, and other data-driven visualizations is D3.js. D3.js was created by Mike Bostock of the New York Times to use HTML, SVG, CSS, and JavaScript to create interactive data-driven graphics. It reads data from HTML and renders it in the way you decide.

D3.js development has taken off since it was first released to the public. The library is very versatile in that it doesn't render just charts and graphs. It provides the building blocks to create charts, graphs, and other interactive graphics that can move and be styled like any HTML element. Even GIS maps in different projections can be shown on a webpage using D3.js and a file format called GeoJSON.

For anyone with experience with jQuery, scripts written with D3.js behave in the same way. You can select HTML elements with the d3.select() or d3.selectAll() methods, which are similar to the jQuery base method. D3 commands can be chained one after another, which is also a favorite feature with many jQuery developers. In the following example, we're using D3 to find elements with the class addflair using the select() method. We then add spans to the elements with related text content:

```
d3.select(".addflair").append("span").text("I've got flair!");
```

Adding the D3.js library with Dojo's AMD

Let's say you want to add D3.js to your mapping application. You find the link to the d3 library, and copy and paste it into your application like so:

```
...
<link rel="stylesheet"
href="https://js.arcgis.com/3.13/esri/css/esri.css" />
<script src="https://js.arcgis.com/3.13/"></script>
<script src=https://d3js.org/d3.v3.js"></script>
</head>
...
```

You cut and paste an example to test if it's going to work. You crank up your browser, and load your page. You wait patiently for everything to load, and it breaks. What happened?

It turns out that the extra libraries loaded after the ArcGIS JavaScript API and interfered with the AMD library references. Let's look at a couple of solutions to load external libraries into an AMD-based application.

Loading another library outside an AMD module

If you are going to work with a JavaScript library outside an AMD module, it's best to load that library before you load the ArcGIS JavaScript API. You would use this if you're adding a map on top of an existing application previously written in another framework.

Loading another library within an AMD module

The other way to handle D3 and other external libraries in your AMD applications is to load them as AMD modules. You can treat them like any other Dojo-based module, and load them into memory only when necessary. This works well with libraries that you use sporadically and don't need at startup. It also works well with libraries that load all the functionality of the library into a single JavaScript object, such as `D3.js` or jQuery.

To load an external library as an AMD module, you must first reference it in `dojoConfig` as a package, just like you did with your custom `Dojo` module in *Chapter 3*, *The Dojo Widget System*. Adding your external library to the packages will tell Dojo's `require()` and `define()` functions where to look for the libraries. Remember that, when listing the location of the library in the package, you reference the file folder of the JavaScript library, not the library directly. For D3, the `dojoConfig` script may look something like the following:

```
dojoConfig = {
  async: true,
  packages: [
    {
      name: "d3",
      location: "http://cdnjs.cloudflare.com/ajax/libs/d3/3.4.12/"
    }
  ]
};
```

Once the library folder reference has been added to your `dojoConfig` variable, you can add it to any `require()` or `define()` statement. Loading the library into an AMD `require()` statement would look like the following:

```
require([…, "d3/d3", …], function (…, d3, …) { … });
```

Using D3.js in our application

In our application, we will explore using D3 to add graphs to our application. We'll use it to replace parts of the `dojox/charting` code where we add the graphs to our map popup. Many of the steps will be similar, but some will be different.

Adding D3.js to the configuration

Since our application relies heavily on the Dojo framework, we will add our `D3.js` library with AMD. We'll add the reference to D3 in our `dojoConfig.packages` list. The new `dojoConfig` script should look like the following:

```
dojoConfig = {
  async: true,
  packages: [
    {
      name: 'y2k',
      location: location.pathname.replace(/\/[^\/]*$/, '') +'/js'
    },
    {
      name: "d3",
      location: "http://cdnjs.cloudflare.com/ajax/libs/d3/3.4.12/"
    }
  ]
};
```

Now that our AMD code knows where to look for the D3 library, we can add a reference to it in our census application. The D3 library will then be available to our census widget, but it will not interfere with other applications that may have their own d3 variable. Our `Census.js` code should look like the following:

```
define([
  ...,
  "esri/config", "d3/d3"
], function (
  ...,
  esriConfig, d3
) {
  ...
});
```

Preparing the popup

D3.js should now be accessible in our widget, and we can prepare the popup to load the data. We're going to set up our code in the same way we did when we loaded the dojox/charting modules. We'll attach the same event to the map.infoWindow object's selection-change event, and on that, we'll run functions to manipulate and render our data.

> Refer to the *Preparing the popup* section in the dojox/ charting portion of the chapter to get the code.

As for the HTML popup templates for the block groups, counties, and states, we can make the same changes we made to the ones in the dojox/charting example. In keeping with best practices on the Internet, we will replace the id tags on the graphing div elements with class tags of the same name (ethnic groups get class="ethnicgraph", for instance). This will cut down on the possibility of HTML id collision. Also, while Dojo widgets require either an HTML element or an id string, D3.js graphs can be added to elements found with any CSS selector.

Processing our data

When we collected the attribute data for the dojox/Charting modules, we had to arrange the attribute data into arrays so they could be consumed by the graphing modules. The same is true for D3.js. We will format the attributes into a list that can be read by the graphs.

Unlike the dojox/charting library, D3.js doesn't have name restrictions on the properties used by the graphing parts. You can give the properties more reasonable names. Functions in D3.js will be added to calculate graph values. Since much of our ethnic, gender, and age data is based on population and sorted by name, it makes sense to name those properties population and name, respectively:

```
...
formatAttributesForGraph: function (attributes, fieldLabels) {
  var data = [], field;
  for (field in fieldLabels) {
    data.push({
      name: fieldLabels[field],
      population: +attributes[field]
    });
  }
  return data;
},
...
```

We add the property names and the population values to a list in the `formatAttributesForGraph()` method. That list will be graphed at a later time. We don't need to change any of the code because we're using the same function to process the attribute data in the `ethnicData()`, `genderData()`, and `ageData()` function.

Displaying the results

Now that we've created our lists of data, we can display them in the graphs on our popups.

Displaying the ethnic graph

For our ethnic graph, we're going to create a pie chart:

1. We'll scale it to fit within a 240 pixel by 210 pixel area in our popup window.

2. We'll add our own color scale with a list of CSS colors.

3. We'll look for the HTML DOM element where we want to put our graph (`class="ethnicgraph"`), and then attach the pie chart graphic.

4. We'll apply the color, size it with our population data, and then label it with the names of the ethnic groups:

```
ethnicGraph: function (data) {
    var width = 240,
      height = 210,
      radius = Math.min(width, height) / 2;

    var color = d3.scale.ordinal()
      .range(["#98abc5", "#8a89a6", "#7b6888", "#6b486b",
      "#a05d56", "#d0743c", "#ff8c00", "#c7d223"]);
    var arc = d3.svg.arc()
      .outerRadius(radius - 10)
      .innerRadius(0);

    var pie = d3.layout.pie()
      .sort(null)
      .value(function(d) { return d.population; });

    var svg = d3.select(".censusethnic").append("svg")
      .attr("width", width)
      .attr("height", height)
      .append("g")
```

```
    .attr("transform", "translate(" + width/2 + "," +
    height/2 + ")");

if (!data || !data.length) {
  return;
}

var g = svg.selectAll(".arc")
  .data(pie(data))
  .enter().append("g")
  .attr("class", "arc");

g.append("path")
  .attr("d", arc)
  .style("fill", function(d) { return
  color(d.data.name); });

g.append("text")
  .attr("transform", function(d) { return
  "translate(" + arc.centroid(d) + ")"; })
  .attr("dy", ".35em")
  .style("text-anchor", "middle")
  .text(function(d) { return d.data.name; });
},
```

When the application draws the graph, it should look something like the
following graph:

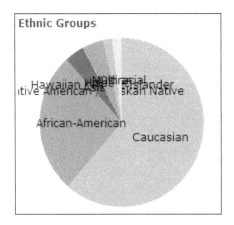

Displaying the gender graph

For the gender graph, we'll start by copying and pasting the code from the ethnic graph. The code is similar, except for two minor changes:

1. For our first change, we'll add custom colors for the male and female populations. Look for where the color variable is assigned, and insert the two color hexadecimal numbers into the color range:

    ```
    genderGraph: function (data) {
    …
      var color = d3.scale.ordinal().range(["#8888ff",
      "#ff8888"]);
    …
    },
    ```

2. Next, we would like to make the labels show both the gender in question, and the actual population. To make a two-line label, we need to add another `tspan` to fill in with the population. We also need to move that label so that it is under the other label, and doesn't cross over it:

    ```
    g.append("text")
      .attr("transform", function(d) { return "translate(" +
      arc.centroid(d) + ")"; })
      .attr("dy", ".35em")
      .style("text-anchor", "middle")
      .text(function(d) { return d.data.name; })
      .append("tspan")
      .text(function(d) { return d.data.population;})
      .attr("x", "0").attr("dy", '15');
    ```

3. Once we run the application and test it with some data, the graph should look like the following image, pending data:

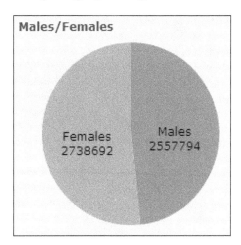

Displaying the age graph

The age graph creates bar graphs from simple html `div` elements. It resizes them according to the data we've provided. We need to calculate a maximum value for the data, so that we can fit the data values within a maximum width. From there, we can draw and label our graphs with the data provided:

```
ageGraph: function (data) {
  // calculate max data value
  var maxData = d3.max(arrayUtils.map(data, function (item)
  {return item.population;}));
  // create a scale to convert data value to bar width
  var x = d3.scale.linear()
            .domain([0, maxData])
            .range([0, 240]);
  // draw bar graph and label it.
  d3.select(".censusages")
    .selectAll("div")
    .data(data)
      .enter().append("div")
      .style("width", function(d) { return x(d.population) + "px";
      })
      .text(function(d) { return d.age + ": " + d.population; });
}
```

Using CSS styling, we can transform the appearance of the data however we wish. In this example, we decided to go with an alternating color theme, using the CSS3 `nth-child(even)` pseudo class selector. You can add your own CSS hover effects to match what we did with `dojox/charting`:

```
.censusages > div {
  background: #12af12;
  color: white;
  font-size: 0.9em;
  line-height: 1.5;
}
.censusages > div:nth-child(even) {
  background: #64ff64;
  color: #222;
  margin: 1px 0;
}
```

Using CSS and our data, we were able to create the following graph:

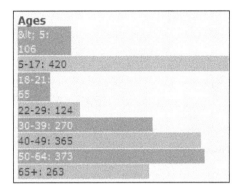

If you would like more information on the D3.js library, there is a wealth of information available. The official D3.js website is at http://d3js.org/. You can go there to find examples, tutorials, and other eye-popping graphics. You can also check out *Data Visualization with d3.js* by Swizec Tellor, *Data Visualization with D3.js Cookbook* by Nick Qi Zhu, and *Mastering D3.js* by Pablo Navarro Castillo.

Summary

Both dojox/charting and D3.js have their advantages and disadvantages in our web mapping applications. The dojox/charting library comes with the ArcGIS JavaScript API, and is easily integrated with existing applications. It provides many themes that can be added quickly. On the other hand, D3.js works with HTML elements and CSS styling to create eye-popping effects. It offers more data visualization techniques than dojox/charting and offers customizable appearances using CSS styling. Your final choice may come down to your comfort level with these tools and your imagination.

In this chapter, we have learned how to incorporate graphs and charts in to our ArcGIS JavaScript API applications. We used graphics libraries provided by the Dojo framework, which created graphics based on data from map features. We also used D3.js to render charts and graphs in our application. In the process, we learned how to load and access other libraries in Dojo's AMD-based applications.

In the next chapter, we'll explore how to mix our ArcGIS JavaScript API applications with other popular JavaScript frameworks.

7
Plays Well with Others

JavaScript development has matured over recent years. Once a simple language used to script a few animations and user interactions, JavaScript applications now support a full range of tasks that were only possible with desktop applications a few years ago. There are numerous well-documented libraries and frameworks that make complicated applications manageable through a web browser.

No JavaScript library or framework does everything perfectly. Most libraries focus on improving a few key requirements for application development, such as DOM manipulation, data binding, animation, or graphics. Some focus on speed; others focus on cross-browser compatibility.

A successful JavaScript developer doesn't have to build an application from scratch using one library or raw JavaScript to make it work. By adding current libraries that work well together, the developer can speed up development and handle multiple browser types without having to worry about them.

In this chapter, we'll cover the following:

- Integrating jQuery into the applications
- Integrating Backbone.js into the applications
- Integrating Knockout.js into the applications
- Discuss Using AngularJS in a mapping application

Compatibility with other libraries and frameworks

One way developers describe the usability of JavaScript libraries is how compatible they are with other libraries. The more libraries your library plays well with, the more likely it can be used on projects. When libraries don't play well with other libraries and frameworks, it can lead to unintended code bugs and actions that nobody saw coming. Since a developer cannot control what other libraries are used in a project with their library, it is better if a library plays well with others.

There are several ways that libraries can conflict with one another. One way that was more common in the past is through namespace collision. If two libraries create the same `cut()` function in the global namespace, you can't be sure which one will be called. Another, more subtle conflict arises through manipulating the prototype of standard JavaScript types. One library will expect the `map()` method called on an array to perform a specific task, but if another library overwrote `Array.prototype.map()`, the results may not turn out as expected, and may break the application. Typically, it's only polite to manipulate the prototype of base JavaScript types if you are patching support for older browsers.

A more recent way that JavaScript libraries can clash is in how they implement modules. Before ECMAScript 6 came out with the standard for modules, there were two schools of module development. One revolved around using CommonJS to define modules, and the other was AMD. Typically, modules from one can be loaded by the other, as they are called and constructed differently. Most small libraries that aren't defined one way or another can be loaded into either one, with a little work.

Even within AMD, there can be some conflicts. Some libraries and frameworks use Require.JS to load their modules. While syntactily similar, there are some differences in older versions of Dojo that create errors when mixing with Require.JS modules. It's something to be aware of when mixing and mashing up JavaScript libraries in your application.

Popular libraries to play with

Now that we've looked at library compatibility as a whole, it's time to check out how the ArcGIS API for JavaScript plays with other JavaScript libraries and frameworks. We'll look at some of the popular libraries used by developers, and attempt to load them along with the ArcGIS API for JavaScript. We'll create the same application using jQuery, Backbone.js, Knockout.js, and AngularJS, and see how they compare. Finally, we'll look at a few do's and don'ts with any framework and the ArcGIS API.

Our story continues

So, our census data app is a big hit for the Y2K society. They like the look of the graphics, and want us to expand it. The functionality of the widget is great, but they've made a new request. It seems that one of the members watched a late night comedian on television the other night, and found out that there are people out there who can't find their state or county on a map.

After discussing possible solutions with our client, we've decided to add some drop-down controls that let the user select a state, county, and even a census block group. On picking a state, the map zooms to the state and shows the data in a popup. Also, the county dropdown is populated with the county names in that state. When the county is selected, the county is selected with a popup, and the block group names are populated for that county and state. Finally, if the block group is selected, the block group is zoomed to, and the data is displayed on the map.

In somewhat related news, we've picked up an intern. He's an eager young kid who likes to follow the latest in JavaScript. He wants to contribute something to this project, and this looks like a good task to let him cut his teeth on.

The problem is, our intern wants to write the code using another library or framework, and not just a simple one. He wants to bring in a number of popular libraries including jQuery, Backbone, Knockout, Node.js, and Angular. While we appreciate his desire to learn (and to pad his resume), we have to have a serious talk with him about picking libraries appropriate for the project.

After talking the intern out of a Node.js project (since this is a client-side app, with no need for server-side work at this point), we explain the general use of the other libraries, he had on his list. We decide to give the intern a learning project, and let him build four copies of the app, each incorporating either jQuery, Backbone, Knockout, or Angular. He can evaluate each app and let us know which worked best with the application. We figure it's okay that he does four times the work, since he's not a paid intern.

We should go ahead and set up four project folders with copies of the D3.js app from our previous chapter's work. In each of the projects, we need to modify the Census.html template in the /js/template folder to show the drop-down menus. We'll add three drop-down menus with bold labels and selects that will be populated later by the four libraries. It should look like the following:

```
<div class="${baseClass}" style="display: none;">
  <span class="${baseClass}-close" data-dojo-attach-
  event="ondijitclick:hide">X</span>
  <b>Census Data</b><br />
  <p>
```

```
    Click on a location in the United States to view the census
data for that region. You can also use the dropdown tools to
select a State, County, or Blockgroup.
</p>
<div >
  <b>State Selector: </b>
  <select></select>
</div>
<div>
  <b>County Selector: </b>
  <select></select>
</div>
<div>
  <b>Block Group Selector: </b>
  <select></select>
</div>
</div>
```

When the new application loads, the new widget should look something like
the following:

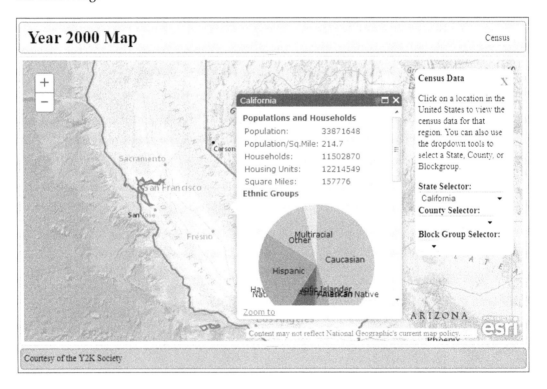

An overview of jQuery

The most popular library for a number of years has to be jQuery. jQuery was originally created by John Resig as a library to handle DOM manipulation and events. Cross-browser compatibility is a strong selling point for this library, since it works in everything from the newest version of Chrome to Internet Explorer versions so old they make most developers cringe. Its intuitive commands and functionality make it easy to pick up for novice developers. The library is so popular that, as of the end of 2013, it is estimated to be used in 61 percent of the top 100,000 websites in the world, according to the jQuery blog (http://blog.jquery. com/2014/01/13/the-state-of-jquery-2014/).

If you understand how jQuery works, you can skip the next section and go to *Adding jQuery in our App*. If not, here's a quick rundown on how jQuery works.

How jQuery works

The jQuery library is incredibly useful for selecting and manipulating DOM elements on a web page. DOM elements can be selected using the jQuery() function, or its common shortcut method $(). It accepts either a DOM element or a string with a CSS selector. To select an element with an id, you would put a # in front of the ID name. To select a CSS class, you insert a . in front of the class name. To select all elements of a tag name, just use the tag name. In fact, almost any legal CSS select statement is supported by jQuery. Here are some examples in the following code:

```
jQuery("#main"); // selects a node with an id="main"
$("#main"); // A common shortcut for jQuery; same as above.
$(".classy"); // selects all elements with class="classy"
$("p"); // selects all paragraphs
$(".article p"); // selects all paragraphs in nodes of
                 // class="article"
$("input[type='text']"); // selects all text inputs
```

Once the nodes are selected by the jQuery object, you can perform a number of functions on them. You can make them appear and disappear using the show() and hide() methods. You can change their style using css(), their properties using attr(), and their content using html() or text() methods. Selected form element values can be retrieved and set using the val(). You can also find elements within the selected elements with find(), and iterate through each item with the each() method. Events can also be assigned using the on() method, and removed using off(). These are just the most common jQuery methods.

Another powerful feature of jQuery is a property called **chaining**. jQuery functions return the jQuery object. You can immediately call another function on what was returned. These chained commands are evaluated from left to right. You can create some powerful transformations on your HTML using this code. For example, if you wanted to turn every unordered list item into a green frog and every ordered list item to a big brown toad, you could do the following:

```
$("ul > li").css("color", "green").text("Frog");
$("ol > li").css({"color": "green", "font-size": "72px"})
   .text("Toad"); // totally legal to do this.
```

Just be warned, as you work with complex chained functions in jQuery, sometimes the target of the function will change. Functions like `each()` and `find()` will change the selection that is passed to the next function. In this example, the developer wants to show a form, set its width to `400px`, clear the values of the text blanks, and turn the labels blue. When they run it, they will find that the labels don't turn blue as they expected:

```
$("form").show().css("width", "400px")
   .find('input').val('')
   .find('label').css('color', '#0f0');
```

When the developer called `find('input')`, it changed the selected items from form elements to input elements. Then, when the `find('label')` was called, jQuery searched for labels that were inside the input tags. Not finding any, nothing turned blue. Thankfully, the developer remembered that jQuery also provides an `end()` function that returns you to your original selection. The following bit of code works much better:

```
$("form").show().css("width", "400px")
   .find('input').val('').end()
   .find('label').css('color', '#0f0');
```

Now that we've taken a crash course in jQuery, let's try to use it in our application.

Adding jQuery to your app

So, our intern is starting to work with the jQuery-based app when he runs into his first problem. How does he make jQuery and Dojo play well together? For the answer to that, we can look at some of the lessons from *Chapter 6, Charting Your Progress*, when we added `D3.js` to our application.

Loading a copy of our `D3.js` app, we'll start by adding a reference to the jQuery library in our `dojoConfig` packages. Remember that it is the JavaScript configuration object that is loaded before we load the ArcGIS Javascript API. We'll add a package with the name and file location, but we'll also add a `main` property that will make the application load the jQuery library. You can download a copy of the jQuery library and place it in an accessible folder in your application, or you can make a reference to an external **content delivery network** (or **CDN**) that's hosting the library. Using an external reference to jQuery hosted by Google, it should look something like the following code:

```
dojoConfig = {
  async: true,
  packages: [
    {
     name: 'y2k',
     location: location.pathname.replace(/\/[^\/]*$/, '') + '/js'
    }, {
     name: "d3",
     location: "http://cdnjs.cloudflare.com/ajax/libs/d3/3.4.12/"
    }, {
     name: "jquery",
     location:"http://ajax.googleapis.com/ajax/libs/jquery/1.11.2",
     main: "jquery"
    }
  ]
};
```

Now that we can access jQuery through AMD, let's open our `Census.js` file and add it as a module in our file. We'll add a reference to it in the `require` statement, and assign it to the module `$`, as it is commonly referred to in most jQuery examples. It should look something like the following code:

```
require([…, "d3/d3", "jquery"], function (…, d3, $) {
  …
});
```

Next, we'll need to update the `Census` widget template to give jQuery something to search for. While updating `Census.html` in the `js/templates/` folder, we'll add a `stateselect`, `countyselect`, and `blockgroupselect` class to each of the `select` menus. While adding IDs would make selecting the elements faster, adding class names will ensure that there is no ID name collision with other widgets in the application. The template will look like the following example:

```
...
<p>
Click on a location in the United States to view the census
data for that region. You can also use the dropdown tools to
select a State, County, or Blockgroup.
</p>
<div >
  <b>State Selector: </b>
  <select class="stateselect"></select>
</div>
<div>
  <b>County Selector: </b>
  <select class="countyselect"></select>
</div>
<div>
  <b>Block Group Selector: </b>
  <select class="blockgroupselect"></select>
</div>
...
```

Since we have something for jQuery to select, we need to let jQuery select it in our code. Add some jQuery selectors to the constructor of our `Census` widget to get the nodes that have the `stateselect`, `countyselect`, and `blockgroupselect` classes, and assign them to the `stateSelect`, `countySelect`, and `blockGroupSelect` properties of our widget, in that order. This is referred to as **caching** our selections, which is a good practice for jQuery apps, since repeating DOM searches can take a long time, especially in a larger application:

```
constructor: function (options, srcNodeRef) {
  ...
  this.map = options.map || null;
  this.domNode = srcRefNode;

  this.stateSelect = $(".stateselect");
  this.countySelect = $(".countyselect");
  this.blockGroupSelect = $(".blockgroupselect");
  ...
},
```

If we ran the application now, we would find that there was nothing in the `stateSelect`, `countySelect`, and `blockGroupSelect` properties. Why? If you remember, back in *Chapter 3, The Dojo Widget System,* we talked about the lifecycle of a `Dojo` widget. We discussed the fact that, while the constructor function runs, the template won't have loaded yet. In fact, it will not be available until the widget runs the `postCreate()` method. We'll need to add `postCreate()` to our application, add a line that refers to the `postCreate` function inherited from the `WidgetBase` class, and then cut and paste the jQuery code we wrote earlier into this function. It should look something like the following code:

```
constructor: function (options, srcNodeRef) {
  ...
  this.domNode = srcRefNode;
  /* stateSelect, countySelect, and blockGroupSelect removed */
  ...
},

postCreate: function () {
  this.inherited(arguments);

  this.stateSelect = $(".stateselect");
  this.countySelect = $(".countyselect");
  this.blockGroupSelect = $(".blockgroupselect");
},
```

Now, when the `postCreate` method is called, `stateSelect`, `countySelect`, and `blockGroupSelect` will be filled with appropriate objects.

Filling in our dropdowns

We need data to fill in our select dropdowns. For that, we'll need to get state, county, and block group names from the map services and fill them in the dropdowns. For the fifty or more states, that is easy, but what about counties and block groups? There are over 250 counties in the state of Texas alone, and even more block groups. We need a systematic way to populate the dropdowns.

What if, every time we selected a state, the county dropdown was filled with all the counties in that state? Also, what if the block groups list wasn't populated until the user selected the county of interest? We can achieve this if we listen to the `select` element's `change` event. When the event fires, we'll have a new selection result from the previous level query for the new list.

We'll start by stubbing out two new methods in the `Census` module, called `queryDropdownData()` and `populateDropdown()`. Instead of adding a list of parameters that we don't know yet, we'll add a single argument called `args` to the `queryDropdownData()` method. To the `populateDropdown()` method, we'll add the `args` parameter, plus a `featureSet` parameter that will come from querying the map data:

```
...
getDropDownData: function (args) { },
populateDropdown: function (args, featureSet) { },
...
```

Adding the QueryTask

Inside the `queryDropdownData()` method, we're going to query a map service to get the list of counties in a state, or block groups in a county. For this, we'll need to add the ArcGIS JavaScript API's `QueryTask`. If you remember from *Chapter 2, Digging into the API*, `QueryTask` lets us pull SQL-like queries to single layers in a map service. We'll need to add references to the `require` statement for the `QueryTask` and its associated task parameter builder, the `Query`. We'll go ahead and construct these in our `getDropdown()` method:

```
require([…, "esri/tasks/Query", "esri/tasks/QueryTask", …],
  function (…, Query, QueryTask, …) {
    …
  queryDropdownData: function (args) {
    var queryTask = new QueryTask(/* url */),
        query = new Query();
  },
    …
});
```

Now we need to define the parameters for both `QueryTask` and the query. For that, we'll use parameters passed through the `args` object. We can define an `args.url` to send a URL string to the `QueryTask`. We can also use `args.field` to set the field name for the data we want returned in the query object, and `args.where` to supply a `where` clause to filter the results. `queryDropdownData()` should now look like the following:

```
queryDropdownData: function (args) {
  var queryTask = new QueryTask(args.url),
  query = new Query();
  query.returnGeometry = false;
  query.outFields = args.fields;
  query.orderByFields = args.fields;
```

```
    query.where = args.where;
    query.returnDistinctValues = true;

    return queryTask.execute(query);
},
```

For our `populateDropdown()` method, we'll take the jQuery-based selector and add the `featureSet` features that will be returned from the `queryDropdownData()` method. Remember that a `featureSet` contains a `features` parameter, which contains a list of graphic results. Out of the graphics returned, we're only interested in the attributes of the field returned. It should look as follows. We're going to use the jQuery `each()` function on the list of features to iterate through, create an option for each result, attach it to the selector, and fill it in with values. It should look like the following:

```
_populateDropdown: function (args, featureSet) {
    args.selector.empty();
    $.each(featureSet.features, function () {
        args.selector.append($("<option />")
            .val(this.attributes[args.fieldName])
            .text(this.attributes[args.fieldName])
            .attr("data-fips", this.attributes[args.fips]));
    });
},
```

Other helper functions

With the addition of `QueryTask`, we can now directly query the state, county, or block group we select from the drop-down menu. We need to define the function that will collect that information from the server. We can call the method in our widget `queryShapeAndData()`, and it will take a single `args` parameter with the data we need from it:

```
queryShapeAndData: function (args) {
    var queryTask = new QueryTask(args.url),
        query = new Query();
    query.returnGeometry = true;
    query.outFields = ["*"];
    query.outSpatialReference = this.map.spatialReference;
    query.where = args.where;

    return queryTask.execute(query);
},
```

While it is possible to add these graphics directly to the map, we should format the data so that it brings up an `infoWindow` as if we clicked it ourselves. For this, we'll add an `_onQueryComplete()` method. It will accept a `featureSet` returned from a `QueryTask`, and return a list of features with appropriate popup templates just like the ones we assign when we identify through clicks. Unfortunately, `featureSets` do not return the same descriptive information as `IdentifyResults`, so we will have to manually add the title of the feature for it to accurately pick the right `InfoTemplate`:

```
_onQueryComplete: function (title, featureSet) {
  featureSet.features = arrayUtils.map(
    featureSet.features,
    lang.hitch(this, function (feature) {
      return this._processFeature(feature, title);
    }));
},
```

Going back to the `queryShapeAndData()` method, we can add the `callback` function to the `execute` statement and have it return a processed result every time. The last part of `queryShapeAndData()` will look like the following:

```
...
return queryTask.execute(query)
  .addCallback(lang.hitch(
    this, this._onQueryComplete, args.title)
  );
},
```

Finally, we need a way to show the queried graphics on the map. We'll create a method called `_updateInfoWindowFromQuery()` that takes a `featureSet`, zooms to its features, and adds the `infoWindow`. We're going to use the `esri/graphicsUtils` module to collect the overall extent of the graphics, so that we can zoom to it. Once the asynchronous zoom finishes, we'll set the graphic on the `infoWindow` and show it. You can see the code that does all this in the following:

```
_updateInfoWindowFromQuery: function (results) {
  var resultExtent =
  graphicsUtils.graphicsExtent(results.features);

  this.map.setExtent(resultExtent)
    .then(lang.hitch(this, function () {
      this.map.infoWindow.setFeatures(results.features);
      this.map.infoWindow.show(resultExtent.getCenter());
    }));
},
```

Handling events

Now, we will add event listeners to the `stateSelect`, `countySelect`, and `blockGroupSelect` items. We'll use the helper functions we developed in previous sections to populate the data, using some of our knowledge of `dojo/Deferred` to connect them asynchronously. Let's begin with the states.

When you select a state from the drop-down menu, the `select` element will fire a `change` event. We're not going to collect data on this event. Instead, we'll get data directly from the drop-down selectors and use that to generate the queries we need.

Let's go ahead and stub out the `_stateSelectChanged()` method in the `Census` widget. It takes no arguments. We'll do the same with the `_countySelectChanged()` and `_blockGroupSelectChanged()` methods as well. Then, using jQuery's `.on()` method, we'll listen for the `change` events in the `stateSelect`, `countySelect`, and `blockGroupSelect` controls. We'll use dojo's `lang.hitch()` method to make sure that, when we say `this`, we mean `this` widget. It should look like the following:

```
postCreate: function () {
  this.stateSelect = $(".stateselect")
    .on("change", lang.hitch(this, this._stateSelectChanged));
  this.countySelect = $(".countyselect")
    .on("change", lang.hitch(this, this._countySelectChanged));
  this.blockGroupSelect = $(".blockgroupselect")
    .on("change",lang.hitch(this, this._blockGroupSelectChanged));
},
…
_stateSelectChanged: function () {},
_countySelectChanged: function () {},
_blockGroupSelectChanged: function () {},
…
```

Handling the state

In the `_stateSelectChanged()` method, we'll start by collecting the name of the state selected in `stateSelect`. If there is a value, we'll start by querying for the graphical data on that state. We'll use the `queryShapeAndData()` method to query for the shape data and process it. When that is complete, we can pass it along to the `_updateInfoWindowFromQuery()` method:

```
_stateSelectChanged: function () {
  var stateName = this.stateSelect.val();
  if (value && value.length > 0) {
    this.queryShapeAndData({
      title: "states",
```

```
        url:
        "http://sampleserver6.arcgisonline.com/arcgis/rest/
        services/Census/MapServer/3",
        where: "STATE_NAME = '" + stateName + "'"
    }).then(lang.hitch(this, this._updateInfoWindowFromQuery),
        function (err) { console.log(err); });
  }
},
...
```

Now that we're showing the graphic, it's time to fill in the county name. We'll use the queryDropdownData() method to query for the list of counties, and using .then() asynchronously, pass the results to our _populateDropdown() method. We'll assign the county names first in the list of queried values, because we want them in alphabetical order. We'll tack that on at the end of the _stateSelectChanged() method, and it should appear like the following:

```
...
// search for counties in that state.
this.queryDropdownData({
  url:
  "http://sampleserver6.arcgisonline.com/arcgis/rest/
  services/Census/MapServer/2",
  fields: ["NAME", "CNTY_FIPS"],
  where: "STATE_NAME = '" + stateName + "'",
}).then(lang.hitch(this, this._populateDropdown, {
  selector: this.countySelect,
  fieldName: "NAME",
  fips: "CNTY_FIPS"
}));
...
```

How the counties differ

The counties should load similarly to the states. The big issue is that we are required to query for the underlying block groups in a county with state and county FIPs codes. There are no state or county names available in the block group census data.

Since we assigned the FIP codes to the data-fips attributes of the drop-down options, there should be a way we can get them, right? Yes we can, but we'll have to take advantage of jQuery's chaining methods. From stateSelect, for instance, we can use the jQuery.find(":selected") method to find the selected option inside the select. From there, we can call jQuery's attr() method to grab the data-fips attribute. For the state and county names and FIPs codes, it should look like the following:

```
_countySelectChanged: function () {
  var stateName = this.stateSelect.val(),
    stateFIPS = this.stateSelect.find(":selected")
      .attr("data-fips"),
  countyName = this.countySelect.val(),
  countyFIPS = this.countySelect.find(":selected")
    .attr("data-fips");
...
```

From here, we can use `stateName` and `countyName` to query for the correct county to show, and `stateFIPS` and `countyFIPS` to get the list of block groups. They'll use the same functions as the `_stateSelectChanged()` method, but with different map services and `where` clauses.

Finally, the block groups

The `_blockGroupSelectChanged()` method is much easier to write, because we're only interested in showing the block group. The key here is to collect the selected state FIP code, the selected county FIP code, and the `blockgroup` value from their respective dropdowns. Then we'll piece together the `where` clause on the query, and request the graphics for the map. The method should look like the following code:

```
_blockGroupSelectChanged: function () {
  var stateFIPS = this.stateSelect.find(":selected").attr("data-
  fips"),
    countyFIPS = this.countySelect.find(":selected").attr("data-
  fips"),
    blockGroup = this.blockGroupSelect.val();
  this.queryShapeAndData({
    title: "Census Block Group",
    url:
    "http://sampleserver6.arcgisonline.com/arcgis/rest/
    services/Census/MapServer/1",
    where: "STATE_FIPS = '" + stateFIPS + "' AND CNTY_FIPS = '" +
    countyFIPS + "' AND BLKGRP = '" + blockGroup + "'"
  }).then(
    lang.hitch(this, this._updateInfoWindowFromQuery),
    function (err) { console.log(err); });
},
```

Filling in the states

Now that the application is wired, we can finally load the state data from the map service. We'll start by querying the state layer for state names and state FIP codes. To get all the states, we'll use a little SQL trick. SQL will return all rows where the `where` clause is true, so if you want all the rows, you have to return something that's always true. In this case, we'll assign `1=1` to the where clause for `queryDropdownData()`, because it is always true that `1=1`.

Once we receive the query results from the server, we'll pass them to our `_populateDropdown()` method. We'll assign the options to the `stateSelect` drop-down, show the state names on each choice, and store the FIP codes in the options as well. We'll add the following snippet to the end of our `postCreate()` method:

```
this.queryDropdownData({
  url:
  "http://sampleserver6.arcgisonline.com/arcgis/rest/
  services/Census/MapServer/3",
  fields: ["STATE_NAME", "STATE_FIPS"],
  where: "1=1",
}).then(lang.hitch(this, this._populateDropdown, {
  selector: this.stateSelect,
  fieldName: "STATE_NAME",
  fips: "STATE_FIPS"
}));
```

So, if we've wired everything up correctly, we can view the site in the browser and open the `Census` widget. We should then see the states already loaded in the drop-down menu. You will have to select a state before the counties fill in, and select a county for the block groups.

Keep this project handy. We'll make copies of this one as a starting point for the other projects. Now let's try to build a Backbone app.

An overview of Backbone.js

With a simple application like the one we wrote with jQuery, the logic can turn into spaghetti-code rather quickly. We need to implement good practices to organize our code. Which practices do we implement? One of the first libraries to try to answer this question is Backbone.js.

Backbone.js is one of the earliest JavaScript libraries to implement a **Model View Controller (MVC)** architecture. MVC organizes code by separating data and business logic (the Model) from the output (the View) and updates both through a separate component (the Controller). With MVC, you don't write a complicated JavaScript function that gets input from some text blanks, adds the contents together, and saves the results. Those three actions can be written into three different functions, separated by what they do and how they fit into the Model, View, and Controller classifications.

Backbone.js requires a couple of other libraries in order to work properly. Backbone uses jQuery to handle DOM manipulation, including showing and hiding elements on the page. Backbone also requires another library called Underscore.js. This library provides a number of functions and shortcuts for dealing with JavaScript objects, arrays, and so forth. Underscore.js provides methods found in dojo/_base/lang and dojo/_base/array, as well as other methods that can help you pull out relevant data from your feature graphics.

Components of Backbone

Compared to most popular libraries and frameworks, Backbone.js is rather simple. It organizes your code into five categories: Model, View, Collection, Event, and Router. These work together to showcase your data and react to user input. All but the events are created through Backbone's extend() method, which contains a JavaScript object that defines the Model, View, Collection, or Router. Events, on the other hand, are defined through the creation of the other items. Let's review each item individually.

The Model

Backbone.js' Model provides a description of the data you're going to use. You can assign default properties, methods, and events to the model that will be called by application features created in the other categories. Models are created through the Backbone.Model.extend() constructor method. Models created with this method become constructors for model data in your application. Data created through models have different methods to get() and set() data, test for the presence of data through has(), and even detect changes in the data through isChanged() or changedAttributes().

Here's an example of a playing card model, as well as a card created using the model. The `CardModel` variable includes `rank` and `suit` properties, as well as a function to describe the card in a single string:

```
var CardModel = Backbone.Model.extend({
  defaults: {
    "rank": "2",
    "suit": "heart"
  },
  description: function () {
    return this.rank + " of " + this.suit + "s";
  }
});
var myCard = new CardModel({rank: "9", suit: "spade"});
myCard.description(); // returns "9 of spades";
```

The View

The Backbone `View` sets up how the data will be presented in the application. The `View` defines the HTML output through a series of parameters defined in the `Backbone.View.extend()` constructor. You can create a view on a particular DOM element in your HTML by assigning the `.el` property in the `extend()` constructor method. You can also assign a `tagName` property to define the HTML element that is created, a `template` property if the content is more complicated than a single element, and even a `className` property to assign a CSS class to the main element.

The view makes heavy use of both jQuery and Underscore.js. For example, while the element of the view is defined by the view's `.el` property, a jQuery version is available by referring to the Views `$el` property. Also, HTML content can be defined through the view's template, which is created through Underscore's `.template()` creator.

When a view is first created, it starts with a method called `initialize()` that you define. In the `initialize()` method, you can assign event listeners to other parts of the view, including models and collections. You can also tell the view to `render()`, or write out the HTML code. The `render()` method that you define is used to add the custom HTML within the `View` element. You can also render other views within the `View`.

In the following code, you can find a sample `View` to show a card:

```
var CardView = Backbone.View.extend({
  tagName: "li"
  className: "card"
  template: _.template("<%= description %> <button>Discard</button>"),
  initialize: function () {
    this.render();
  }
  render: function () {
    this.$el.html(
      this.template(this.model.toJSON())
    );
  }
});
```

The Collection

If a `Backbone` model describes one piece of data, how do you describe a list of them? That's where a `Collection` comes in. A collection represents a list of data items of a particular model. As is probably no surprise, a `Collection` constructor can be created using the `Backbone.Collection.extend()` method.

Collections offer a number of methods for managing the content of your list. You can `.add()` or `.remove()` model-defined items from your collection list, as well as `.reset()` the entire list to whatever you pass as a parameter. You can define a `url` parameter as a JSON data source, and `.fetch()` the data.

In the following code, you can see how a `Collection` is created using a deck of cards. It is based on the `CardModel` defined in the model:

```
Var CardCollection = Backbone.Collection.extend({
  model: CardModel,
  url: "http://cardjson.com", // made up for this example
  deal: function () {
    return this.shift();
  }
});
var deck = new CardCollection();
deck.fetch(); //load cards from website if it existed.
```

Implementing a router

Backbone routers help define the state of the application through the URL. They respond to changes in the URL hash, or the text following the # symbol in the URL. The hash was originally created in web pages to allow you to click an anchor tag to move down the page to related content, without reloading the entire page. When a Backbone router is enabled, you can change the hash, say, through a button or anchor click, and it will run some JavaScript code in response to the content. All this happens, and the page doesn't reload.

This lack of page reloads when the router changes makes single-page applications possible. Instead of loading new pages, Backbone can simply show different views in response to the router. This gives a snappier response on the page.

Handling events

Events are defined within the other items created by Backbone. Event listeners are attached to elements of the Model, View, Collection, or Router, by way of the on() method. The on() method takes three parameters, a string containing the name of the event, the function that is to be called when the event occurs, and the context that defines what this is.

Events within HTML elements created by the view are defined in a different way. The Backbone.View.extend() constructor contains an events parameter that is defined by an unusually formatted JSON object. Event names, and jQuery selectors for the elements, are used as the key, and a string containing the name of the function called in the view makes up the key value. Example events might look like the following:

```
...
events: {
  "keypress input": "stopOnEnter",
  "click #mybutton": "itsMine",
  "focus .tinything": "bigScare",
  "blur .tinything": "makeSmallAgain"
}
...
```

Putting some Backbone in your app

Since Backbone uses jQuery under the hood to work with the DOM, we can reuse much of the code from our jQuery application. We'll be using the same ArcGIS JavaScript API modules to interact with the map services. We'll only change how the drop-down options are rendered and how the change events on those dropdowns are handled. So, let's start by making a copy of our jQuery application and naming the folder `Backbone`.

Next, we'll need to add references to the Backbone and Underscore libraries in our `dojoConfig`, so that they are available through AMD. We'll load them from a CDN source for this application, although you're free to download them into folders for your own applications:

```
dojoConfig = {
  async: true,
  packages: [
    ...
    {
      name: "jquery",
      location:
      "http://ajax.googleapis.com/ajax/libs/jquery/1.11.2",
      main: "jquery"
    }, {
      name: "underscore",
      location:
      "http://cdnjs.cloudflare.com/ajax/libs/underscore.js/1.7.0",
      main: "underscore"
    }, {
      name: "backbone",
      location:
      "http://cdnjs.cloudflare.com/ajax/libs/backbone.js/1.1.2",
      main: "backbone"
    }
  ]
};
```

After that, we'll make reference to the jQuery, Underscore, and Backbone libraries in our `define` statement in the `Census.js` file. The files should load in as so:

```
define([…, "d3/d3", "jquery", "underscore", "backbone"],
  function (…, d3, $, _, Backbone), {
    …
});
```

Defining the models

Now, we have the opportunity to define the data models that we'll be working with. If you remember from the jQuery exercise, we were primarily interested in the names and the FIP codes for the census locations. In the `postCreate()` method of the census widget, we'll define our models using default values from those fields:

```
postCreate: function () {
  …
  // Backbone Models
  var State = Backbone.Model.extend({
  defaults: {
    "STATE_NAME": "",
    "STATE_FIPS": ""
  }
  });

  var County = Backbone.Model.extend({
  defaults: {
    "STATE_NAME": "",
    "STATE_FIPS": "",
    "NAME": "",
    "CNTY_FIPS": ""
  }
  });

  var BlockGroup = Backbone.Model.extend({
  defaults: {
    "BLKGRP": "0"
  }
  });
},
```

Defining the collections

For the state, county, and block group collections, we'll simply define them based on the corresponding models we defined previously. We'll then create collection objects that are tied to the Dojo `dijit`. It should look like the following:

```
postCreate: function () {

  // Backbone Collections
  var StateCollection = Backbone.Collection.extend({
    model: State
  });
  var CountyCollection = Backbone.Collection.extend({
    model: County
  });
  var BlockGroupCollection = Backbone.Collection.extend({
    model: BlockGroup
  });

  this.stateCollection = new StateCollection([]);
  this.countyCollection = new CountyCollection([]);
  this.blockGroupCollection = new BlockGroupCollection([]);
},
```

Defining the views

Now that we've defined our models and collections, it's time to create a view for them. Our view should create the options with the data that we need in our application. We'll need to create a separate view for each of the dropdowns, since assigning templates when the view is created causes errors with this version of the Backbone.

Let's start with the `StateView` variable. The `StateView` variable will be created through `Backbone.View.extend`. In the `StateView` variable, we want to define a template, an `initialize()` method, and a `render()` method. The `initialize()` method will listen for the collection's `reset` event, and cause it to `render()` again. The template is defined by Underscore's `_.template` function called on an HTML string pulled up by jQuery. The jQuery selector will look for our state option template by looking for an element with the ID of `stateitemtemplate`:

```
// Backbone Views
var StateView = Backbone.View.extend({
  initialize: function () {
    this.collection.on("reset", this.render, this);
  },
```

```
      template: _.template( $("#stateitemtemplate").html()),
      render: function () {
        // compile the template using underscore
        var template = this.template,
          el = this.$el.empty();
        // load the compiled HTML into the Backbone "el"ement
        _.each(this.collection.models, function (item) {
          el.append(template(item.toJSON()));
        });
      }
    });
```

In the preceding view, the render function does two things. First, it loads the view template and the empty jQuery selection object into variables. Next, it iterates over each of the collection models using Underscore's each() method, fills in the template with the JSON content from the model, and appends it inside the select element. Some other Backbone examples would stop the option creation from appending the options to the select element in a separate view, but this method was chosen for compact purposes.

Now that the StateView is defined, you can copy and paste the code and tweak it to create separate CountyView and BlockGroupView constructors. In each of those, the only thing you need to change is the template jQuery selector, to #countyitemtemplate and #blkgrpitemtemplate, respectively. Keep the initialize() and render() methods the same:

```
    ...
    var CountyView = Backbone.View.extend({
      ...
      template: _.template( $("#countyitemtemplate").html()),
      ...
    });

    var BlockGroupView = Backbone.View.extend({
      ...
      template: _.template( $("#blkgrpitemtemplate").html()),
      ...
    });
```

Finally, we'll assign our actual view properties to these `View` constructors. With each view, we'll assign the element they'll be rendered from, which are the `select` elements in our widget and the collections they will be using:

```
this.stateView = new StateView({
  el: $(".stateselect"),
  collection: this.stateCollection
});

this.countyView = new CountyView({
  el: $(".countyselect"),
  collection: this.countyCollection
});

this.blockGroupView = new BlockGroupView({
  el: $(".blockgroupselect"),
  collection: this.blockGroupCollection
});
```

Creating templates

Thinking back to the jQuery app, we filled in the select dropdowns with options tags that contained data on both the name of the feature and a FIP code with which we could query the next level down. We need to create the same HTML elements in HTML templates rather than piecing them together through code. We'll do this by adding HTML templates to our main page.

How do we create HTML templates that aren't visible on the page? We can do this by inserting them into script tags. We'll give each template script tag a type of `text/template`. This lets us know that the content of the script is actually HTML code. Browsers will look at the type, not know what to do with a `text/template` type, and simply ignore it.

So, let's create templates for the state options, county options, and block group options. We'll assign the templates IDs of `stateitemtemplate`, `countyitemtemplate`, and `blkgrpitemtemplate`, as we added in our code. In each template, we'll assign the value, text, and `data-fips` value to the appropriate item in the model. Take a look at the following templates:

```
<script type="text/template" id="stateitemtemplate">
  <option value="<%= STATE_NAME %>" data-fips="<%= STATE_FIPS %>">
    <%= STATE_NAME %>
  </option>
</script>
```

```html
<script type="text/template" id="countyitemtemplate">
  <option value="<%= NAME %>" data-fips="<%= CNTY_FIPS %>">
    <%= NAME %>
  </option>
</script>
<script type="text/template" id="blkgrpitemtemplate">
  <option value="<%= BLKGRP %>" data-fips="<%= BLKGRP %>">
    <%= BLKGRP %>
  </option>
</script>
```

Depending on the template library you use, different libraries have different ways to assign values. Underscore's `template()` method wraps the attribute names in `<%= %>` tags. You can use other template libraries, such as Handlebars.js, but, since Underscore is required, why not use what we have?

Wiring events

Now let's make things happen. We're going to reuse the event listeners we created for the jQuery exercise, and make them work for Backbone. We'll start by looking at the `_stateSelectChanged()` method.

The first thing that changes is how we collect `stateName` and `stateFIPS`. Instead of referring the `stateSelect` that was previously defined as a jQuery object, we'll access the `select` through the `stateView.$el` property. Remember that, in a `View`, the `el` property exposes the DOM element, while the `$el` property exposes the jQuery element. For the other selection change listeners, we can find and replace `countySelect` and `blockGroupSelect` with `countyView.$el` and `blockGroupView.$el`, respectively.

The only other part that needs changing is how the new drop-down data is populated after it is queried form the map service. We can replace the `_populateDropdown()` method with a simple anonymous function. In the anonymous function, we'll create a list of feature attributes from the `featureSet` using Underscore's `pluck()` method. It goes item by item through an array, and grabs the property of the item you describe and puts that in a new list. Next, we'll assign that list to the `countyCollection` through its `reset()` method. That's all that is needed to update the county list. The same process can be performed on the `_countySelectChanged()` method to repopulate the block groups. Your code changes should look like the following:

```
_stateSelectChanged: function () {
  var stateName = this.stateView.$el.val();
  …
  This.queryDropdownData({
    …
  }).then(lang.hitch(this, function (featureSet) {
    this.countyCollection.reset(
      _.pluck(featureSet.features, "attributes")
    );
  });
},

_countySelectChanged: function () {
  var stateValue = this.stateView.$el.val(),
    stateFIPS = this.stateView.$el.find(":selected")
      .attr("data-fips"),
    countyValue = this.countyView.$el.val(),
    countyFIPS = this.countyView.$el.find(":selected")
      .attr("data-fips");
  …
  This.queryDropdownData({
    …
  }).then(lang.hitch(this, function (featureSet) {
    this.blockGroupCollection.reset(
      _.pluck(featureSet.features, "attributes")
    );
  });
},

_blockGroupSelectChanged: function () {
  var stateFIPS = this.stateView.$el.find(":selected")
      .attr("data-fips"),
    countyFIPS = this.countyView.$el.find(":selected")
      .attr("data-fips"),
    blockGroup = this.blockGroupView.$el.val();
  …
},
```

Getting the Backbone to dance

Finally, we need to populate the initial value of `stateView`. We'll use the `queryDropdownData()` method call at the end of jQuery's `postCreate()`. If we make the same changes to this call that we made to the event listeners, we should be able to populate the state drop-down menu. From there, we should be able to populate the other menus through the event listeners we assigned to the `View` elements:

```
postCreate: function () {
  …
  this.queryDropdownData({
    url:
    "http://sampleserver6.arcgisonline.com/arcgis/rest/
    services/Census/MapServer/3",
    fields: ["STATE_NAME", "STATE_FIPS"],
    where: "1=1",
  }).then(lang.hitch(this, function (featureSet) {
    this.stateCollection.reset(
        _.pluck(featureSet.features, "attributes")
    );
  }));
```

 If you want to learn more about implementing ArcGIS JavaScript APIs using Backbone.js and Marionette, you can review the blog posts by Dave Bouwman on the matter at `http://blog.davebouwman.com/2013/02/20/part-1-app-design-and-page-layout/`. For more information on Backbone.js, you can read *Backbone.js Patterns and Best Practices* by Swarnendu De.

An overview of Knockout.js

Another JavaScript library that can be used to create an interactive single page application is Knockout.js. Knockout was developed by Steve Sanderson of Microsoft, though it's not considered a Microsoft product. It's based on Windows Presentation Framework in that it uses the **Model-View-ViewModel (MVVM)** architecture and allows two-way binding on observed properties. Two-way binding means that data isn't just written to an HTML element, but it can also be updated, like a text input field in a form, and the data will be already reflected in the application.

Knockout and MVVM

The MVVM is similar in nature to MVC architecture. Both use a model to get to the data, and a `View` to show the data. However, instead of an active controller directing the model and views, the `ViewModel` sits under the UI layer and exposes functions and data from the model to the `View`. The `ViewModel` typically knows nothing about the `View` it's working with. It simply stores and provides information as requested by the HTML `View`.

In Knockout, the `ViewModel` is created like any normal JavaScript object, with a few additions from Knockout. Properties of the `ViewModel` that are used in two-way binding are created using Knockout's `observable()` and `observableArray()` constructors. This allows these properties to be accessed by the `View`, and updated without having to update the DOM, as you would have to in jQuery. A constructor looks something like the following:

```
// note that ko is the global library object for knockout.js
function PersonViewModel (firstName, lastName) {
    this.firstName = ko.observable(firstName);
    this.lastName = ko.observable(lastName);
}
```

The HTML document acts as a `View`, and can be bound to the `ViewModel` through the HTML5-based `data-*` attributes (more specifically, the `data-bind` attribute). When the browser loads an HTML view and a script containing the `ViewModel`, the Knockout will bind attributes in the data-bind tags to the appropriate fields in the `ViewModel`. For the preceding `ViewModel` created, you might see some HTML like the following:

```
<div>
    <label for='fninput'>First Name:</label>
    <input type='text' id='fninput' data-bind='value:firstName' />
    <br />
    <label for='lninput'>Last Name:</label>
    <input type='text' id='lninput' data-bind='value:lastName'/>
    <br />
    <p>
      Hello,
      <span data-bind='text: firstName'></span> 
      <span data-bind='text: lastName></span>!
    </p>
</div>
```

In the `ViewModel`, properties can be added that are computed values based on other observables in the `ViewModel`. These are created with the `ko.computed()` constructor. For example, we can take the `PersonViewModel` in the preceding code and add a computed `fullName` property that automatically updates when the first or last name changes:

```
function PersonViewModel (firstName, lastName) {
  this.firstName = ko.observable(firstName);
  this.lastName = ko.observable(lastName);
  this.fullName = ko.computed(function () {
    return this.firstName() + " " + this.lastName();
  }, this);
}
```

Knockout doesn't have many extra features that other libraries and frameworks possess, such as routers and AJAX requests. It typically relies on other libraries, such as `Sammy.js` for routing and jQuery for AJAX. However, what it does offer is two-way binding that works even in older browsers.

Using Knockout in our app

Let's make another copy of our jQuery app and name the folder `Knockout`. We will not need jQuery for this application, since we can use Knockout and the ArcGIS JavaScript API to handle those functions. We'll start by replacing all the references to jQuery with Knockout references. The `dojoConfig` script at the head of the document should look like the following:

```
dojoConfig = {
  async: true,
  packages: [
    {
      name: 'y2k',
      location: location.pathname.replace(/\/[^\/]*$/, '') + '/js'
    }, {
      name: "d3",
      location: "http://cdnjs.cloudflare.com/ajax/libs/d3/3.4.12/"
    }, {
      name: "knockout",
      location:
      "http://cdnjs.cloudflare.com/ajax/libs/knockout/3.2.0",
      main: "knockout-min"
    }
  ]
};
```

Next, we'll add a reference to Knockout in our `Census.js` file. We'll keep the other AMD modules and code, and we'll replace things as we go. The `define()` statement at the top of `Census.js` should look a bit like the following:

```
define([…, "d3/d3", "knockout"], function (…, d3, ko) {
    …
});
```

Defining the ViewModel

We know what kind of data model we're dealing with, but what we need is a `ViewModel` to organize it. We can create the `ViewModel` constructor in our `Census` dijit, and expose it for use by our widget.

Our `ViewModel` for this widget only requires six items. We need to maintain the lists of states, counties, and block groups, to populate the `select` elements, as well as the selected values. For the `stateList`, `countyList`, and `blockGroupList` properties of the `ViewModel`, we'll construct Knockout's `observableArrays` for each. `selectedState`, `selectedCounty`, and `selectedBlockGroup` will each be created using Knockout observables. You can see how we construct the `ViewModel` in the widget in the following example:

```
…
SelectorViewModel: function () {
    this.stateList = ko.observableArray([]);
    this.selectedState = ko.observable();
    this.countyList = ko.observableArray([]);
    this.selectedCounty = ko.observable();
    this.blockGroupList = ko.observableArray([]);
    this.selectedBlockGroup = ko.observable();
},
…
```

Adding custom binding handlers

In our jQuery and Backbone applications, we attached event listeners to the `select` elements so that, when they change, we can query for census data and populate the next select down. In Knockout, we can do the same thing using custom binding handlers. Binding handlers have two optional methods: `init()` and `update()`. The `init()` method runs when binding first occurs, while the `update()` method runs every time the bound value changes. Both `init()` and `update()` have five arguments, as follows:

- `element`: The HTML DOM element involved in the binding.

- `valueAccessor`: A function that gives access to the observable property bound to the element. To get the value of this property, call the `ko.unwrap()` method on the value returned from `ValueAccessor()`.

- `allBindings`: An object used to get the particular bindings to the element, like the text, value, or name. Binding properties can be retrieved using `allBindings.get()`.

- `viewModel`: This was the old way to get at the entire `ViewModel`, prior to version 3.x.

- `bindingContext`: This is the way to get to all the bindings. `bindingContext` may have at least some of the following:

    - `$data`: The current `ViewModel` assigned to this element

    - `$rawData`: Direct access to the values held in the `ViewModel`

    - `$parent`: Access to the parent `ViewModel` this `ViewModel` may be assigned to

    - `$parents`: An array object providing access to each tier of `ViewModel` connected to this `ViewModel`

    - `$root`: This grants direct access to the base `ViewModel`, originally bound to the entire page

We need to create the binding handlers before we apply bindings to the page. While we can create the binding handlers earlier, we're going to assign them in the `postCreate()` method of our widget, since that's where we've applied the changes in our other applications. We'll start by creating some empty binding handlers for `stateUpdate`, `countyUpdate`, and `blockGroupUpdate`. We're only interested in the `update()` methods, so we'll leave the `init()` method out. The following code is the empty version of `stateUpdate`:

```
...
postCreate: function () {
  ko.bindingHandlers.stateUpdate = {
```

```
    update: function (
      element, valueAccessor, allBindings, viewModel, bindingAccessor
    ) {
      // content will come here shortly.
    }
  };

},
...
```

Within the binding handlers, and the subsequent calls to the ArcGIS JavaScript API modules, the widget scope is going to get lost. In the postCreate() method, we'll create a variable named self, and assign a reference to the widget to it, like so:

```
postCreate: function () {
  var self = this;
  ko.bindingHandlers.stateUpdate = {…};
  ...
},
```

With the binding handlers set, we'll bring our code over from our previous event handlers. From _stateSelectChanged(), we'll copy over the code and make our changes. First, we'll change how the functions collect the state names and census FIP codes for the queries. Once we have the names, getting the visual data will be easy. Finally, we'll change how the ViewModel is updated once the query for the counties in the state is finished. The stateUpdate binding handler should look like the following:

```
...
// within the postCreate() method.
var self = this;

ko.bindingHandlers.stateUpdate = {
  update: function (
    element, valueAccessor, allBindings, viewModel,
    bindingAccessor
  ) {
    // retrieve the selected state data.
    var stateData = ko.unwrap(valueAccessor()),
      stateName, stateFIPS;

      // if stateData is valid and has a state name…
    if (stateData && stateData.STATE_NAME) {
       // assign state name and FIPS code.
      stateName = stateData.STATE_NAME;
      stateFIPS = stateData.STATE_FIPS;
```

```
// query for shape and attributes to show on the map.
// Replace all "this" references with "self" variable.
self.queryShapeAndData({
  url:
  "http://sampleserver6.arcgisonline.com/arcgis/rest/
  services/Census/MapServer/3",
  where: "STATE_NAME = '" + stateName + "'"
}).then(
  lang.hitch(self, self._updateInfoWindowFromQuery),
  function (err) { console.log(err); });

// search for counties in that state.
self.queryDropdownData({
  url:
  "http://sampleserver6.arcgisonline.com/arcgis/rest/
  services/Census/MapServer/2",
  fields: ["NAME","STATE_NAME","CNTY_FIPS","STATE_FIPS"],
  where: "STATE_NAME = '" + stateName + "'",
}).then(function (featureSet) {

  // create an array of county data attributes
  var theCounties = arrayUtils.map(
    featureSet.features,
    function (feature) {
      return feature.attributes;
    });

  // insert a blank value in the beginning of the array.
  theCounties.splice(0, 0, {
    NAME:"",STATE_NAME:"",CNTY_FIPS:"",STATE_FIPS:""});

  // assign the list to the countyList in the ViewModel
  bindingContext.$data.countyList(theCounties);
  });
  }
 }
};
```

For the county and the block groups, we'll follow similar patterns in `countyUpdate` and `blockGroupUpdate`. Remember the following changes for this app:

- Replacing references to `this` with the variable `self`
- Get county and block group feature attribute data from `ko.unwrap(valueAccessor())`
- Collect the list of feature attributes for the drop-down list using the `Dojo` module method `arrayUtils.map()`
- Add an attribute with blank values to the first feature of the feature attributes list
- Add the new list to the `bindingContext.$data` array

Defining the View

Now that we've done all the hard work setting up the `ViewModel` and the related functionality, let's work with some HTML. Open the `Census.html` template in the `js/template/` folder. This is where we are going to apply our `View`. Start by removing the class assignments on the three `select` elements, and replace them with the text `data-bind=""`.

Next, we're going to assign four properties within each data-bind context: `options`, `optionsText`, `value`, and the appropriate binding handler we created in the previous section. The options will be assigned to the appropriate `observableArray` list. The `optionsText` will be the field name that we want to see on the options. Both the value and the binding handler we created will be bound to the selected observable for that type. The `Census.html` file should look like the following:

```html
...
<div >
  <b>State Selector: </b>
  <select data-bind="options: stateList,
                     optionsText: 'STATE_NAME',
                     value: selectedState,
                     stateUpdate: selectedState"></select>
</div>
<div>
  <b>County Selector: </b>
  <select data-bind="options: countyList,
```

```
                                optionsText: 'NAME',
                                value: selectedCounty,
                                countyUpdate: selectedCounty"></select>
    </div>
    <div>
      <b>Block Group Selector: </b>
      <select data-bind="options: blockGroupList,
                                optionsText: 'BLKGRP',
                                value: selectedBlockGroup,
                                blockGroupUpdate: selectedBlockGroup"></select>
    </div>
    ...
```

Applying the ViewModel

Now that we have a working `View` and `ViewModel`, and the code that links it to our `Model`, it's time to put it all together. Once the page has loaded in the `postCreate()` method, and all our binding handlers have been assigned, it's safe to call `ko.applyBindings()` to the `ViewModel`. In most examples that you see online, the `ViewModel` is bound as follows:

```
ko.applyBindings(new ViewModel());
```

It would be great to finish with that and call it a day, but we don't have any state data to start out the `SelectViewModel`. We're going to have it assign it some way. We're going to take some code from the jQuery and Backbone code to assign the initial values to the state dropdowns. We'll then modify it to fit the pattern we established for assigning the other dropdowns. We'll query for a list of states, and add the list to our working `ViewModel`, like so:

```
...
var viewModel = new this.SelectorViewModel();
ko.applyBindings(viewModel);

this.queryDropdownData({
  url:
  "http://sampleserver6.arcgisonline.com/arcgis/rest/
  services/Census/MapServer/3",
  fields: ["STATE_NAME", "STATE_FIPS"],
  where: "1=1",
}).then(function (featureSet) {
  // make a list of the feature attributes of the states
  var theStates = arrayUtils.map(
    featureSet.features,
```

```
    function (feature) { return feature.attributes; });
    // add in a blank value to the beginning of the list
    theStates.splice(0, 0, {STATE_NAME: "", STATE_FIPS: ""});
    // apply the list of states to our viewModel
    viewModel.stateList(theStates);

});
```

 For more information on KnockoutJS, you can visit `http://knockoutjs.com` or `http://learn.knockoutjs.com` for interactive tutorials. For books on the library, you can check out *KnockoutJS Starter* by Eric M. Barnard, or *Mastering KnockoutJS* by Timothy Moran.

A brief overview of AngularJS

One of the more recent popular frameworks that we can use alongside our JavaScript application is AngularJS. Angular was originally created at Google as a language to help designers code. It quickly grew into the JavaScript framework of choice, both at Google and abroad. The core Angular development team makes sure that Angular excels in data binding and testability.

Angular works by parsing through the HTML page for specific element attributes. These attributes give directives to Angular about how to bind input and output elements on the page to JavaScript variables in memory. Data that is bound to the HTML can either be defined in code or requested from a JSON source. The binding is two way, similar to Knockout, but this framework has been fleshed out to provide more popular features found in other libraries.

Angular and MV*

While Angular was designed with MVC in mind, developers argue that it doesn't follow the MVC pattern. They say that the controller aspects don't really behave as controllers, but more like presenters (in MVP) or `ViewModel` (MVVM). This led some Angular developers to coin the term **Model View Whatever** (**MVW** or **MV*** for short). Whatever architecture style it uses, it has caught the attention of many developers. It's currently one of the most popular frameworks, favored even more than jQuery by most JavaScript developers.

Angular vocabulary

One of the barriers to using AngularJS is the new vocabulary it introduces. Many of the terms were created by computer scientists at Google, so they may sound very academic. However, when we place them within the context of a language we know, we can understand the concepts better. We're going to discuss the following Angular terms, and how they relate to our mapping application:

- Controller
- Service
- Directive

The app controller

In Angular, the **Controller** is a JavaScript object connected by two-way binding to an HTML document using Angular tags. Changes to values within the scope of the controller will change the values displayed on the HTML document. The Angular controller is comparable to the `ViewModel` in Knockout. For map-driven applications written with Angular, the map is often defined as part of the main controller.

The app service

In Angular, a **service** is a JavaScript object used to communicate with servers and other data sources. Services are created once, and keep running through the life of the application in the browser. The equivalent item in the ArcGIS JavaScript API would be a task like `QueryTask` or `GeometryService`. In fact, if you wrote an Angular app using the ArcGIS JavaScript API, you could wrap a `QueryTask` within an Angular service and serve data that way.

The app directive

In Angular, a **directive** is a custom element with its own behavior defined in code. The directive loads a template HTML file, applies its own bindings, and displays on a web page. In the ArcGIS JavaScript API, the equivalent to a directive is the Dojo `dijit` that we learned about in *Chapter 3*, *The Dojo Widget System*. The `dijit` defines an HTML template and the JavaScript behavior bound to it. One difference between directives and dijits is that Angular directives are allowed to give an HTML element a custom name, such as `<overviewmap></overviewmap>`, while `dijits` must be built on existing HTML elements. Also, Angular events and descriptions are described with `ng-*` HTML parameters rather than `data-dojo-*` parameters.

Making a web mapping app more Angular

Since AngularJS is a framework, it is expected to handle the work of page layout, event handling, and so on. However, Dojo is also a framework, and has its own opinions about how those page events should be handled. Can the two get along?

The key to using AngularJS and the ArcGIS JavaScript API side by side is to let Angular handle most of the page layout, but use the map, widgets, and task from the ArcGIS JavaScript API to communicate with ArcGIS Server.

Much of the work needed to make our application run with AngularJS would take far too long to explain, and deserves its own chapter. Instead, I've supplied a few resources you can look over to integrate Angular into your own application.

> For more information about incorporating ArcGIS JavaScript API with Angular.JS, you can read blog posts by Rene Rubalcava at `http://odoe.net/blog/using-angularjs-with-arcgis-api-for-javascript/` or read his book *ArcGIS Web Development*, published by Manning Press. You can also review code from ESRI at `https://github.com/Esri/angular-esri-map`. Finally, for more information about Angular.JS, you can read *Instant AngularJS Starter* by Dan Menard, or *Mastering Web Application Development with AngularJS* by Pawel Kozlowski and Peter Bacon Darwin.

Overall results with other frameworks

All the frameworks we have reviewed in this chapter have worked with the ArcGIS API for JavaScript. Some, such as jQuery, slipped right and could be used right away. Others, such as Backbone and Angular, required a significant rewrite of the application. Each of these libraries takes up significant bandwidth when the browser downloads the website. These libraries and frameworks would have to bring something very important to make it worth the wait.

jQuery

The jQuery library performs a number of functions found in Dojo. Both work well in a wide array of browsers. While jQuery's functionality and style make it easier to work with the HTML DOM, its event handling doesn't couple directly with Dojo's event handlers. In other words, `$(map).on("click", …)` doesn't do the same thing as `map.on("click", …)`.

If you're creating an application that's jQuery centric and you want to add an ArcGIS-based map, it's perfectly fine to mix the two libraries together. You can easily define map operations with Dojo and other form operations through jQuery. However, jQuery doesn't add anything to the ArcGIS JavaScript library that can't be accomplished by importing a couple of extra Dojo modules (which saves on download bandwidth as well).

Backbone.js

Backbone organizes your code well, but it requires a lot of code to make it work. All the models, views, and other features have to be defined individually, and coupled with one another. There are other extension libraries, such as `Marionette.js`, that can make Backbone easier to code. Using Backbone could be very beneficial to architect the application around the map, but for this simple job, it was a little excessive.

Knockout and Angular

Both Knockout and Angular are great frameworks for CRUD (Create, Read, Update, and Destroy) applications, but they don't bring anything new and powerful to web mapping applications. They can slow an application down if a lot of two-way binding is applied to an HTML document. Also, since Angular is a framework, it is written to handle many of the user interactions behind the scenes. Large portions of the code would have to be rewritten if Angular was added to an existing ArcGIS JavaScript API application.

In conclusion, we could have easily accomplished what we wanted using Dojo and the ArcGIS JavaScript API. We could save time and bandwidth on a smaller device like a smartphone or tablet by writing these portions using the ArcGIS JavaScript API. However, it helps to know how to incorporate ArcGIS maps with existing applications written in these other frameworks.

Summary

In this chapter, you have learned how to use the ArcGIS JavaScript API along with other JavaScript libraries. We built applications using jQuery, Backbone.js, Knockout. js, and Angular.js. We compared the different uses of the libraries and frameworks, and how they worked with the ArcGIS API and Dojo.

In the next chapter, we'll tackle a topic that strikes fear into the hearts of some JavaScript developers… styling.

8
Styling Your Map

It's the one word that strikes fear into the heart of developers: style. They think it's a field for right-brain dominant artists. They treat color theory, typography, and whitespace as a foreign language. They long to push these tasks onto a web designer and focus on algorithms and API calls.

Many small firms, government agencies, and departments don't have the luxury of having a web designer on their staff, or even on call. These small organizations tend to consist of one or more people hired for their technical and analytical skills, while design skills are left as an afterthought. Maybe you're working for one of those right now.

While this chapter may not turn you into an instant web designer, it will help you use CSS to effectively lay out your web map applications.

In this chapter, we'll cover the following topics:

- How CSS is applied to an HTML document
- The different ways to position the map on a page
- How to use Dojo's `dijit/layout` modules to style your page
- How to add Bootstrap to the layout of your page

The inner workings of CSS

As we mentioned in *Chapter 1*, *Your First Mapping Application*, the **Cascading Style Sheet** (**CSS**) tells the browser how to render an HTML page. As the browser scans through the HTML, it scans through all the applicable CSS styles from CSS files, as well as any overriding styling within the HTML, to see how it should render the element. CSS descriptions, such as color and font size, often cascade down from one element to its children unless specifically overridden. For example, the style applied to the div tag will also apply to the p tags inside it, as shown in the following code:

```
<div>
  <p>I'm not red.</p>
  <div style="color:red;">
    <p>You caught me red-handed.</p>
    <p>Me too.</p>
  </div>
  <p>I'm not red</p>
</div>
```

CSS works a little differently from most programming languages. In JavaScript, when you have a bug in part of your program, you can step through the code in your browser until you find the part that breaks. In CSS, however, the rules that define the look of an element can be stretched across multiple style sheets, and can even be written inside the element. The appearance of the element can also be impacted by elements both inside and out.

Selector specificity

The browser decides how elements should be styled by comparing the selector types used to define the style. It applies weights based on the type and quantity of selectors used to define a style. Browsers have five basic ranks for CSS selectors, based on how specific they are. Technically, there is a zeroth selector when you use the * to select every element on the page but, compared to the other selectors, it has no value. Selector ranks are as follows:

- By element (for example h1, div, or p)
- By element classes, attributes, and pseudo selectors. Some examples include:
    - .email or .form (classes)
    - input[type='text'] (attributes)
    - a:hover or p:first-child (pseudo selectors)
- By IDs (for example #main or #map-div)
- By inline styles ( )
- By styles marked !important

A common way to note specificity is to count the number of selectors in each category and separate them with a comma. A p selector gets a specificity of 1, and a p > a gets a specificity of 2. However, a p.special > a gets a specificity of 1,2, because the class falls in a separate, higher category. A #main selector has a specificity of 1,0,0 while the inline style of a p tag earns a specificity of 1,0,0,0. The powerful !important clause is the only thing that can override an inline selector, and it earns a specificity of 1,0,0,0,0.

When comparing selector specificity, a single higher rank selector beats any number of lower rank selectors. In the case of a tie, the next lowest is compared. For example, let's take a look at the following HTML snippet:

```
<style>
  .addr { background: red; }
  .start.blank { background: orange; }
  .help.blank.addr { background: yellow; }
  #citytxt { background: green; }
  input[type='text'] { background: blue; }
  input { background: purple; }
</style>
...
<input type="text" id="citytxt" class="addr start blank help" />
```

What do you think the background color will be from this? The correct answer is green. The #citytxt rule is a third ranked selector, since it points to an individual element on the page. If we look at the selectors along with their specificity ranks, they look as follows:

```
<style>
  .addr { background: red; }                   /* 0,1,0 */
  .start.blank { background: orange; }         /* 0,2,0 */
  .help.blank.addr { background: yellow; }     /* 0,3,0 */
  #citytxt { background: green; }              /* 1,0,0 */
  input[type='text'] { background: blue; }     /* 0,1,1 */
  input { background: purple; }                /* 0,0,1 */
</style>
```

So what happens when everything else is equal?

Equal selector specificity

When two or more rules have equal selector specificity, the last one listed wins. This is another feature of the cascading effect of CSS. We always put the custom CSS after the Dojo and ArcGIS JavaScript API style sheets in our applications. Any style changes we make will then not be overridden by the other style sheets.

With the last in wins rule, we can undo any unintended side effects of CSS rules applied to our widgets. We don't always have to resort to using `!important` tags or writing inline styles that get lost in our code reviews. We can use the same strength of selector specificity, and get the result we want, as long as we put it after the old rule.

Styling tips and tricks

Now that we have an idea how CSS works, we can build on this logic to create some working styles for our application:

- We'll begin by studying some of the bad patterns we need to avoid. We'll look at how they impact styles and further development negatively.

- Next, we'll look into some good practices, such as using responsive design and normalizing style sheets.

- We'll look at how to organize your style sheets to make them easier to extend and troubleshoot.

- We'll cover how to position your map wherever your application needs it.

Styling don'ts

Before I go too far with telling you what to do, let's go through a few things you should avoid. These bad design habits are typically picked up while working through beginner tutorials and copying and pasting single-page applications off the Internet:

- Styling your elements inline: Trying to change the appearance of 20 or more paragraphs one by one is a pain.

- Making everything important: The important clause allows you to override styles imposed by other widgets and imported style sheets, but don't get carried away.

- Having lots of references to individual IDs: A few element ID references are fine but, if you want to reuse your CSS files on other pages, or in other projects, you want them as general as possible. Not everyone will be using your `#pink_and_purple_striped_address2_input` element.

- Writing new changes to your CSS file at the bottom of your page: We all know that last in wins, but if you slap every new update at the bottom of the page, the file becomes a junk-drawer of unorganized CSS rules.

Like any hard and fast rules, there are appropriate times to break them. But, by working within the bounds of these rules, you make it easier for yourself and others to maintain your applications.

Responsive design

The **responsive design** movement has taken a firm hold on website development. Responsive design revolves around the idea that a website should be useable on a wide variety of screen sizes, from large monitors to mobile phone screens. This reduces the costs of maintaining multiple websites for desktop and mobile users.

On the surface, responsive design involves assigning percentage widths and heights instead of fixed sizes, but there is more. Fluid grid layouts support multiple columns on wider screens while collapsing down to a single column on narrow screens. Images at different resolutions can be served for tablets and screens with retina displays for crisper looking images. CSS media queries can change how elements are displayed at different sizes or across different media.

Maps created with the ArcGIS JavaScript API work well with responsive design layouts. The maps keep track of the size of their HTML container elements, and size changes, while updating their contents. While the map scale stays the same, the extent is recalculated, and new map tiles are requested for locations not previously stored in memory. These changes can be triggered by resizing the browser window in a desktop, or turning a tablet sideways in a mobile browser.

Normalize.css

It can be frustrating trying to make your application look good in your browser. Bringing in more browsers compounds the problem. Many browsers have unique rendering engines on different devices which make the same HTML elements look different on each device. Can't they all come to the same conclusion about how HTML elements should look?

Developers and designers often use a CSS file called `normalize.css` (`http://necolas.github.io/normalize.css/`). This style sheet styles HTML elements so that they all look similar on different browsers and devices. It cuts down on the guesswork when you are concerned about how a page will look.

The `normalize.css` file style is typically inserted as the first style sheet in the HTML document head. Any changes you make to your website's style will then be made after the normalize rules are applied, and are less likely to be overridden. Some CSS frameworks like Twitter Bootstrap, already incorporate `normalize.css` in their style sheets.

Organizing your CSS

As previously discussed in the list of things not to do with your styles, the greatest offenders involve using higher selector specificity than needed, and treating the style sheet like a junk drawer. By organizing your style sheet properly, you can cut down on both of these offenders, and make your application easier to style and maintain. Let's go through a few best practices.

Organizing by selector specificity

Current trends in web design call for CSS selectors to be organized from the lowest selector specificity to the highest. All your `div`, `h1`, and `p` tags might be put at the top of the page in your style sheet. After the appearances of the HTML elements are defined, you add various classes, selectors, and pseudo-selectors to describe how those would change the appearance of your elements. Finally, you could assign the appearances of an individual element by referencing its `id`.

Group by module

Applications written with the ArcGIS JavaScript API can easily be organized by dijits, so why not arrange the style sheets by dijits as well? You can define the styling of individual dijits in the application after you define the style of your page. You can separate your CSS into logical sections using `/* comments */` between the module and dijit styles.

A class for everything

A common practice when organizing code for selector specificity is to assign CSS classes as much as possible. Your map dijits may have an `.open` class that sets the `width` and `height`. The same dijit may have a `.closed` class that hides the dijit. Using the `dojo/dom-class` module, you can add, remove, and toggle the classes you define, however you want:

```
.open {
  width: 30%;
  height: 100%;
  display: block;
}
```

```
.closed {
  width: 0%;
  display: none;
}
```

Using descriptive classes makes it easier to see what your application is doing, especially when you look at the page source. Descriptive classes are easier to reference in your style sheet.

Media queries

Media queries provide effective ways to create custom looks and responsive grids on different screens. You can change the look of your site depending on the media type (screen, print, projector, and so on), the screen width, and even the pixel depth (retina displays versus standard desktop screens).

One thing to consider when organizing your code is that media queries should be placed after normal selectors. You can then take advantage of the last in wins principal, and use the same selectors to show different results when the screen size changes. I've been guilty of not paying attention to where I placed my media queries, and wasted time troubleshooting why my transitions weren't occurring. Only later did I find, in the mess of CSS, where I had applied a rule after my media query that cancelled the effect.

Positioning your map

We can't always rely on other frameworks and style sheets to properly position our maps. Sometimes, we have to get our hands dirty with CSS and do it ourselves. We'll go through some styling scenarios for our maps, and look at what CSS rules we need to apply to the map element to position it properly. All examples assume that you're creating a map on a `div` element with the ID of "map".

The fixed-width map

By default, a map is created with a specific width and height. The width and height can be any non-negative number, from a whole screen, to a narrow column. If a height is not assigned, a default `height` of `400px` is assigned to the map element. You can see a simple, non-responsive CSS style for the map here:

```
#map {
  width: 600px;
  height: 400px;
}
```

Stretching the map to fullscreen

Sometimes, your map is more than just an important piece of the page. Sometimes, the map needs to take up the whole page. That is what this full screen size represents. This style works, assuming that the HTML and body tags have a width and height of 100% as well. This full screen style can also be assigned to a map that is supposed to fill 100% of the area in another element:

```
#map {
  width: 100%;
  height: 100%;
}
```

Floating the map to the side

Sometimes you don't want a full map. Sometimes you just want a small map on the side of the screen, showing the location of whatever content it's sharing the page with. You can then float the content off to the side. Floating an element to the right or the left lets other content fill in around it. This technique is typically used with photos and text, where the text flows around the photo. It works for a map as well:

```
#map {
  width: 30%;
  height: 30%;
  float: right;
}
```

Positioning the map top and center

Sometimes you need to center your map in your layout. You have some text, and you just want the map to line up nicely in the middle. With this centering trick, you can horizontally center any block type element on a page. Here, you set the position to relative, and you assign a right and left margin of auto. The browser will automatically assign an equal number to the right and left margin, essentially centering it. But remember, this must be performed on a block element (like the map) with relative positioning. Remove any of these criteria, and the trick doesn't work, as shown in the following code:

```
#map {
/* just in case something sets the display to something else */
  display: block;
  position: relative;
  margin: 0 auto;
}
```

Covering most of the page with the map

If you need an almost full page effect, where you need to leave room for a title bar or a column to the right or left, you can use absolute positioning to stretch the map. Absolute positioning takes the element outside of the normal layout, and lets you position it wherever you want.

Once you assign absolute positioning to the map, you can use the top, bottom, left, and right values to stretch the map out. By assigning a value of 0 to the bottom, you're telling the browser to set the bottom edge of the element at the bottom of the page. By assigning a value of 40px to the top, you are telling the browser to assign the top of the map element 40 pixels from the top of the page. By assigning both left and right values, you are stretching the map element out in both directions.

As a caveat, remember that absolutely positioned elements escape the bounds of their location, and will be positioned either on the whole page, or inside the first parent element that has relative positioning:

```
#map {
  position: absolute;
  top: 40px; /* for a 40px tall titlebar */
  bottom: 0;
  left: 0;
  right: 0;
}
```

Centering a map of known width and height

Sometimes, you need to put your map in the center of the page. You want to create a modal effect, where the map is centered both vertically and horizontally, kind of like a modal dialog. If you know the width and height of the map, you can pull this off easily. You assign absolute positioning to the map, and set the top and left edges of the map at 50% of the page. This won't look right at first, until you assign the margins. The trick is to assign negative top and left margins, with values that are half the height and width of the map elements, respectively. What you get is a vertically and horizontally centered map that also works in older browsers:

```
#map {
  width: 640px;
  height: 480px;
  position: absolute;
  top: 50%;
  left: 50%;
  margin: -240px 0 0 -320px;
}
```

Centering a map of unknown width and height

If you implement percentages or other units into the styling of your map, you may not know at any one time how wide the map is. We can use absolute positioning to put the upper left corner of the element in the middle of the page, but how can the element be shifted so that it sits in the middle of the page? There is an easy way to center the map both vertically and horizontally when the width and height is variable. It requires a CSS3 transformation.

We can translate or move the element in any direction by using the CSS3 transformation. The first value moves it to the right or left, while the second value moves it up and down. Negative values signify translation to the left and up. We can apply width and height in pixels, or we can apply a percentage of the element's width to center it:

```
#map {
  position: absolute;
  top: 50%;
  left: 50%;
  width: 60%;  /* any width is okay */
  height: 60%; /* any height is okay */
  -webkit-transform: translate(-50%, -50%);
  -moz-transform: translate(-50%, -50%);
  -ms-transform: translate(-50%, -50%);
  -o-transform: translate(-50%, -50%);
  transform: translate(-50%, -50%);
    /* transform shifts it halfway back*/
}
```

 CSS3 transformations are available in most modern browsers, with some slightly older ones requiring vendor prefixes. Internet Explorer 8 and earlier do not support these transformations. See http://caniuse.com/#feat=transforms2d for more details.

Troubleshooting

With the rising popularity of web development tools for browsers, your browser is the best tool for troubleshooting styling issues on a page. Mozilla Firefox initially had the most advanced inspection tools using a free third-party add-on called **Firebug**. Later, Chrome released its own development tools, while Firefox and Internet Explorer eventually built and improved their own. All modern desktop browsers now provide advanced JavaScript and CSS information for desktop and mobile devices.

Most browser developer tools can be opened using the same keyboard shortcuts. Internet Explorer responds to the *F12* key as far back as version 8. Chrome and Firefox also respond to *F12*, with the keyboard combination of *Ctrl + Shift + I* (*Cmd + Opt + I* for a Mac).

Responsive resizers

All desktop browsers shrink and grow as you maximize and shrink them. Many modern browsers, however, have extra features and add-ons that can help you test your applications as if they were mobile browsers. The latest editions of the Firefox Developer Edition have a **Responsive Design View** that resizes your browser depending on the mobile device. It can rotate the screen when the user rotates their phone, and even triggers touch events. Google Chrome has a **Device Mode** that lets you select from popular smartphones and tablets, and can simulate slower network speeds, as well as pretending to be a mobile browser by changing the user agent it sends on requests. The latest versions of Internet Explorer also have these in their developer tools.

Now that we've reviewed the tools we can use to test our layouts, let's look at the tools the ArcGIS JavaScript API provides to lay out our applications.

Dojo layout

Dojo uses its own framework to control the layout of the application. Dojo's layouts can be found in the `dijit/layout` modules. These modules can be used to implement full page applications with all their features implemented, by using the Dojo framework.

Dijits created with `dijit/layout` modules can be encoded directly in the HTML. These are encoded using the `data-dojo-type` attribute. Properties for these, including styling and behavior, are encoded in the `data-dojo-props` attribute of the element. These dijits can be loaded from HTML by using the `dojo/parser` module to parse the HTML.

Applications where the `dijit/layout` elements are loaded through HTML often break when `dojo/parser` doesn't have access to a `layout` module. Make sure that all the modules for the layout elements used in the HTML have been loaded in the `require()` or `define()` statement that calls the `parse()` method of the `dojo/parser` module. Check for misspellings, either in the module loaders, or in the HTML `data-dojo-type` attributes.

Layouts created with the `dijit/layout` modules can be divided into two classifications: **Containers** and **Panes**. Pane elements are generally located inside containers.

Containers

Containers are parent elements that control the position and visibility of child panes assigned within them. Containers come in a variety of shapes and functions. Some can show multiple panes at once, while others show one or a few at a time. In JavaScript, if you have access to the Container `dijit`, you can access the pane elements inside it by calling the container's `getChildren()` method.

Let's look at a few of the common containers.

LayoutContainer

The `LayoutContainer` allow other panes to be positioned around a central pane. The center pane in a `LayoutContainer` is assigned a region attribute of center. Panes surrounding it are assigned region values of `top`, `bottom`, `right`, `left`, `leading`, or `trailing` to define their position in relation to the center pane. Multiple panes can have the same region attribute, such as two or three left panes. These will stack side by side to the left of the center pane.

BorderContainer

The `BorderContainer` is derived from the `LayoutContainer`. As the name implies, it adds borders to section off the different panes in the application. `BorderContainers` can also provide `livesplitters`, which are draggable elements that let the user resize panes as they see fit.

AccordionContainer

The `AccordionContainer` arrange and transition between panes using an accordion effectIn this arrangement, pane titles appear stacked on top of one another, and only one pane is visible at any time. The contents of the other panes are hidden by the accordion effect. When the user selects another pane within the `AccordionContainer`, the panes animate in a sliding motion, hiding the current pane and showing the selected pane.

TabContainer

The TabContainer wi provides a tab-based organization of content. Tabs with the titles of the ContentPane enclosed describe the content, and clicking on those tabs removes the content visibility. The effect is similar to a Rolodex or a file folder, where you flip through tabs to view the content you need.

Panes

Panes in the dijit/layout modules provide a container in your application in which to put user controls and widgets. You can write HTML content or add other dijits. Panes can be encoded in HTML, or created with JavaScript and attached to their parent container. Let's look at a couple of the panes available in Dojo using the ArcGIS JavaScript API.

ContentPane

The ContentPane tile is the most common pane inserted in a container. It can be inserted as a pane inside all the other containers, with the exception of the AccordionContainer. On the surface, they appear to be glorified div elements that also track size and relationship to other dijits around them. But a ContentPane tile can also downloads and displays content from other web pages. Setting the ContentPane tile's href property will download and display another web page's HTML content in a single pane in your application.

AccordionPane

One or more AccordionPane panes are added within an AccordionContainer to display their content in a collapsible format. AccordionPane titles are stacked on top of each other and, as you click the titles, the content slides into view, covering the previously open AccordionPane. An AccordionPane otherwise exhibits the same functional behavior as a ContentPane.

Now that we've reviewed how the Dojo framework handles the application layout, let's look at using an alternative style framework.

Bootstrap

If you're looking for an alternative to Dojo's way of laying things out, you might consider Bootstrap. Bootstrap is a popular styling framework originally created by developers at Twitter. The story goes that the developers needed a way to release websites quickly, so they drafted a set of style sheets as a starting point for their projects. The styling templates proved very popular because they were easy to use and met the needs of most web developers. The template, originally named Twitter Blueprint, was later released in August 2011 as Bootstrap.

Bootstrap provides developers with responsive design styles that work well on both desktop and mobile browsers. Responsive grids can be fine-tuned to give you multi-column layouts that collapse to smaller sizes in smaller browser windows. Bootstrap provides stylish looking form elements, and buttons big enough for fat fingers on phone browsers. The framework provides easy to understand CSS classes, and the documentation and style sheets provide guidance on how to use the framework. From picture icons that can be understood across language barriers, to JavaScript plugins that create modal dialogs, tabs, carousels, and other elements we're used to on websites. Entire applications can be created using Bootstrap.

While Bootstrap styling doesn't require JavaScript libraries, all of the JavaScript plugins require jQuery to run. That is not very helpful for those of us using Dojo, but we do have an alternative.

ESRI-Bootstrap

Combining Bootstrap's ease of use and the compatibility with the ArcGIS JavaScript API, ESRI created ESRI-Bootstrap library (`https://github.com/Esri/bootstrap-map-js`). The style framework resizes maps and other elements just like Bootstrap, while many of the elements retain the same look and feel of a Bootstrap site. Dialogs don't run off the screen and elements respond to both mouse clicks and touches. Finally, ESRI Bootstrap can be used with either a combination of Dojo and jQuery, or Dojo alone.

We are going to add ESRI-Bootstrap on top of our Y2K application. You can use the `Dojo/jQuery` application we wrote in *Chapter 7, Plays Well with Others*. We're going to use a pure Dojo application written in parallel with jQuery to show how you can add ESRI-Bootstrap to an application without jQuery.

Restyling our app

We recently let our intern create multiple copies of our application using other frameworks. While they were busy doing that, we decided to write the update using just the ArcGIS JavaScript API. The Y2K society checked out our app and approved our changes, but that wasn't all they had to say.

When we met with the Y2K society board, we found a critic in a new member of the board. They thought it looked okay, but it needed a more modern look. When asked for details, they showed us websites on their tablet that they thought looked good. All the sites we were shown had one thing in common, they were all built using Bootstrap. He managed to convince others on the board that our app needed to embrace this new style.

Going back to the drawing board, we looked at what ESRI had to offer. We decided to incorporate ESRI-Bootstrap into our application. It provides a modern feel.

Adding ESRI-Bootstrap to our app

Let's start by adding references to ESRI-Bootstrap to our application. The ESRI-Bootstrap **Getting Started** page tells us to download a ZIP file of the latest version published on GitHub. Click on the **Download ZIP** button on the right hand side of the GitHub page. Once downloaded, unzip the file.

We're primarily interested in the `bootstrapmap.js` and `bootstrapmap.css` files in the `src/js/` and `src/css/` folders, respectively. The rest of the files contain templates and samples that you can look at to see how they use Bootstrap with their maps. Copy these files into the `js` and `css` folders.

Next, we'll add the necessary references to the head tag of our `index.html` file. We can remove the `nihilo.css` external style sheet as we're no longer using the Dojo widgets to lay out our application, and add the following:

```
...
<link href="http://netdna.bootstrapcdn.com/bootstrap/3.2.0/css/
  bootstrap.min.css" rel="stylesheet" media="screen" />
<link rel="stylesheet"
  href="https://js.arcgis.com/3.13/esri/css/esri.css" />
<link rel="stylesheet" type="text/css" href="css/bootstrapmap.css" >
<link rel="stylesheet" href="css/y2k.css" />
...
```

Now that we have added the stylesheets, we need to add a reference to the Dojo version of bootstrap within our `dojoConfig` variable. The Dojo Bootstrap library, which covers features normally handled by jQuery, can be found at `https://github.com/xsokev/Dojo-Bootstrap`, but we'll pull it from `https://rawgit.com`, which serves GitHub code directly. It should look as follows:

```
dojoConfig = {
    async: true,
  packages: [
    …
    {
      name: "d3",
      location: "http://cdnjs.cloudflare.com/ajax/libs/d3/3.4.12/"
    },
    {
      name: "bootstrap",
      location: "http://rawgit.com/xsokev/Dojo-Bootstrap/master"
    }
  ]
};
```

Now that the head has been taken care of, it's time to overhaul the body.

Bootstrapping our HTML

Since ESRI-Bootstrap provides plenty of CSS styling and JavaScript, most of the work involves overhauling our HTML. We can start by removing the references to the Dojo layout widgets in the body of our main HTML page. You can remove the `class="nihilo"` from the `body` element, the `dijit/layout/BorderContainer` from the `#mainwindow` div, and all the `dijit/layout/ContentPane` references.

Replacing the header with a navbar

Our first job is to replace the header in the application. There are many ways you could start off a page like this but we'll use the Bootstrap `NavBar` in place of the header since the initial layout of the page was a full page app with no scrolling, You can see several examples of how to implement a `navbar` in HTML at `http://getbootstrap.com`.

We'll replace the header `div` with a `nav` element with the classes of `navbar` and `navbar-default` in the header. We'll add a `div` with the class of `container-fluid` inside that, since the size and appearance of its contents varies with different browsers. We'll add two `div` elements to the `container-fluid` div element. The first will have a class of `navbar-header`, and the second will have the classes of collapse and `navbar-collapse`.

We'll add a button to the `navbar-header` that will only appear when the browser is narrow enough. The button will have the classes of `navbar-toggle` and `collapsed`, plus a `data-toggle` attribute of collapse. This button will be floated to the far right when visible. The title of the map will go in a span next to the toggle button. This span will be given a class of `navbar-brand`.

We'll add the button that shows our `Census` dijit to the `navbar-collapse` div element. We'll add the `btn` classes (to make it a button), `btn-default` (to make it a default color), and `navbar-btn` (to make the button style fit in the `navbar`). When we are done, the HTML should look as follows:

```
...
<body>
  <nav class="navbar navbar-default">
      <div class="container-fluid">
      <div class="navbar-header">
        <button type="button" class="navbar-toggle collapsed"
         data-toggle="collapse"
         data-target="#bs-example-navbar-collapse-1">
          <span class="sr-only">Toggle navigation</span>
          <span class="icon-bar"></span>
          <span class="icon-bar"></span>
          <span class="icon-bar"></span>
        </button>
        <span class="navbar-brand" >Year 2000 Map</span>
      </div>
      <div class="collapse navbar-collapse"
         id="bs-example-navbar-collapse-1">
         <button id="census-btn"
             class="btn btn-default navbar-btn">Census</button>
      </div>
      </div>
  </nav>
...
```

Rearranging the map

Now that the header has been redesigned, it's time to look at the rest of the page. We will want to pull the census widget location out of the map in order to incorporate a Bootstrap modal dialog. We can move the footer inside the map to make up the space lost in previous versions. We can define the map and related items below the header, as follows:

```
...
<div class="modal fade" id="census-widget"></div>
<div id="map">
```

```
    <span id="footer">Courtesy of the Y2K Society</span>
</div>
<script type="text/javascript" src="js/app.js"></script>
...
```

Restyling our app

We'll need to manually restyle the page since we're removing all traces of Dojo's layout system. We can remove most of the styling references since they'll clash with the Bootstrap style. We'll keep the basic styling for the HTML and body tags, since they make the application a fullpage app. We'll also keep the styling for the D3.js graphs, but we can delete the rest of the styles.

We need to stretch the map from top to bottom and from end to end. If we don't, the map will be limited in its width and height. We can use absolute positioning to stretch the map across the page. We'll use the almost full page style we talked about earlier. Since the toolbar on the page is 50 pixels high (which you'll see when you experiment with the app), we'll set the top of the map to be 50px from the top. The bottom, right, and left sides can positioned on the edges of the screen:

```css
#map {
  position: absolute;
  top: 50px; /* 50 pixels from the top */
  bottom: 0; /* all the way to the bottom */
  left: 0; /* left side 0 units from the left edge */
  right: 0; /* right side 0 units from the right edge */
}
```

The other item we'll need to restyle is our footer element. We can position it at the bottom of the page, using the same technique we used on the map. We can also make the background semi-transparent, for a nice effect:

```css
#footer {
  position: absolute;
  bottom: 5px; /* 5 pixels from the bottom */
  left: 5px; /* left side 5 pixels from the left edge */
  padding: 3px 5px;
  background: rgba(255,255,255,0.5); /* white, semitransparent */
}
```

Once these styles have been applied, we can see a working example of our map. You can see an example of the map in the following picture:

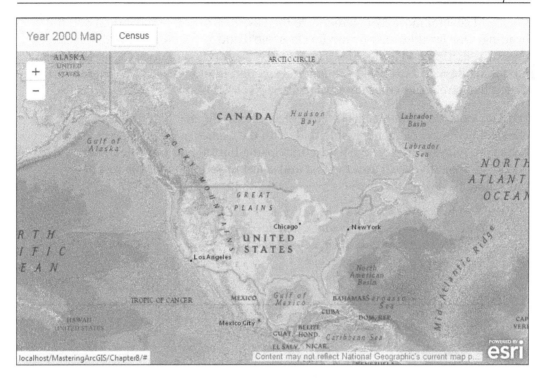

Making our Census dijit modal

Now that our page has been transformed into a Bootstrap application, we need to add the same to our Census dijit. We need to tap into Bootstrap's modal widgets to imitate the effect of our floating dialog.

Open the Census.html file in the js/templates/ folder. In the base div element, add the classes modal and fade. The modal class tells Bootstrap that this will be a modal dialog, while fade describes how the element will hide and show. We'll also add a data-backdrop attribute to the element and set it to static. This will create the generic modal dialog that keeps the rest of the page from being clicked while it's open. In this case, we'll abandon the idea that closing the widget will turn off the map's click events:

```
<div class="${baseClass} modal fade"
   style="display: none;" data-backdrop="static">
...
</div>
```

We'll add several more `div` elements to the base `div` to define the modal and the heading. One level inside our modal class, we'll add a `div` element with the classes of `modal-dialog` and `modal-sm`. The `modal-dialog` class defines the style for the modal, while `modal-sm` makes the modal smaller. Removing `modal-sm` creates a dialog that stretches across larger screens.

We'll create a `div` with the class `modal-content` in the `div` with the `modal-dialog` class, and two `div` elements inside that, with the classes of `modal-header` and `modal-body`, respectively. We'll add our closing button and our title to the `modal-header` div. We'll add the text and select dropdowns for the rest of our dijit to the `modal-body` div:

```
<div class="${baseClass} modal fade"
   style="display: none;" data-backdrop="static">
   <div class="modal-dialog modal-sm">
     <div class="modal-content">
       <div class="modal-header">
       ...
       </div>
       <div class="modal-body">
       ...
       </div>
     </div>
   </div>
</div>
```

We'll replace the `dijit` closing event with Bootstrap's modal closing event in the `modal-header` div. We'll add a class of `close` to the button, and a `data-dismiss` attribute of modal. Most Bootstrap examples also include ARIA attributes to handle screen readers and other accessibility tools, so we'll add an `aria-hidden` value of `true` so that screen-readers do not read aloud the *X* that we place in that. For the title, we'll surround the `Census` data in a span with the class of `modal-title`. It should look like the following code:

```
...
<div class="modal-header">
  <button type="button" class="close"
    data-dismiss="modal" aria-hidden="true">x</button>
  <span class="modal-title">Census Data</span>
</div>
...
```

We'll add our description paragraph to the modal-body div and format the select elements so that they appear as form elements. We'll add the class form-group to the div elements that surround our select dropdowns. This lines up the content and adds proper spacing and formatting. We'll replace the b tags with label tags, and add the control-label class. We'll add the form-control class to the select element. This stretches the select dropdown across the width of the dialog. Our HTML should look as follows:

```
...
<div class="modal-body">
  <p>
    Click on a location in the United States to view
    the census data for that region. You can also use
    the dropdown tools to select a State, County, or
    Blockgroup.
  </p>
  <div class="form-group" >
    <label class="control-label">State Selector: </label>
    <select class="form-control"
      data-dojo-attach-point='stateSelect'
      data-dojo-attach-event='change:_stateSelectChanged'>
    </select>
  </div>
  <div class="form-group">
    <label class="control-label">County Selector: </label>
     <select class="form-control"
      data-dojo-attach-point='countySelect'
      data-dojo-attach-event='change:_countySelectChanged'>
     </select>
  </div>
  <div class="form-group">
    <label class="control-label">
      Block Group Selector: </label>
    <select class="form-control"
      data-dojo-attach-point='blockGroupSelect'
      data-dojo-attach-event='change:_blockGroupSelectChanged'>
    </select>
  </div>
</div>
...
```

We'll add a `div` with the class of `modal-footer` to our `modal-body` div. Here, we can add another button to dismiss the dialog, just in case the user doesn't notice the *x* in the upper corner. We'll format the dismiss button by adding the classes `btn` and `btn-default`, affecting the shape and color, in that order. We'll also add our `data-dismiss` attribute and set it equal to `modal`. It should look as follows:

```
...
<div class="modal-footer">
  <button id="btnDismiss" type="button"
    class="btn btn-default" data-dismiss="modal">
    Dismiss
  </button>
</div>
</div>
...
```

Once the widget HTML is properly formatted, the application should look like the following image. Notice the dropdowns and buttons with wide spacing, which makes it easier to click on them on a smaller mobile device:

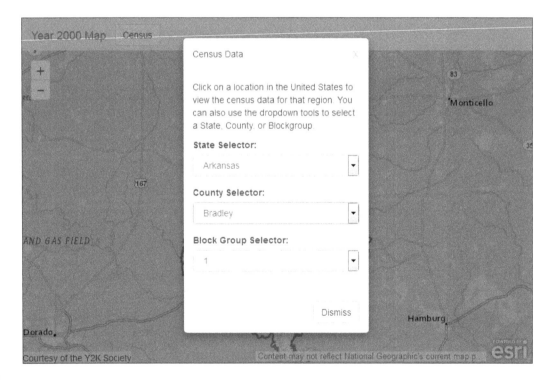

Now that the application is functionally complete, we can modify parts of it to create our own look and feel. Remember, Bootstrap was meant to be a starting point for creating a website. It doesn't have to be the end result. We can still change colors and other features to make the application our own.

Summary

In this chapter, we've discussed the different ways in which you can style your ArcGIS JavaScript API application. We looked at how CSS works, how rules affect each other, and how your browser decides which CSS rules to follow. We looked at the Dojo layout modules, and how those can be used to handle the appearance and functionality of your application. We also looked at ESRI-Bootstrap, a version of Bootstrap that can run with the ArcGIS JavaScript API. Finally, we added the ESRI-Bootstrap look to our application.

In the next chapter, we'll be going mobile. We'll create a mobile application that works on most tablets and phone browsers.

Mobile Development

9

Mobile web development has caused quite a stir in the last few years. When Apple introduced the iPhone, it didn't support third party plugins such as Flash and Silverlight. This challenged web developers to deliver worthwhile web experiences on the mobile platform with only HTML, CSS, and JavaScript. Proposed feature enhancements such as HTML5, CSS3, and ECMAScript 5 and 6, along with more powerful browsers, have improved the mobile browsing experience.

Companies and organizations have taken different approaches to delivering the mobile web experience. Some organizations reroute mobile browsers to sites that serve mobile content only (with URLs such as `mobile.example.com` or `m.example.com` instead of `www.example.com`). Others have used responsive design and mobile first strategies to deliver the same content, formatted differently, for phones, tablets, and desktop screens. Still others, such as Facebook, have given up on mobile web development and focused on mobile apps, using native applications or hybrid web apps.

In this chapter, we'll look at the following:

- What makes developing for mobile devices different from desktop website development
- How to use the ArcGIS compact build
- How to control the mobile user experience with `dojox/mobile` modules

Embracing mobile

Mobile web development is a fast-growing market. In India in 2012, the percentage of Internet content served from mobile devices surpassed that of desktop computers. In the US, reports show that web traffic from mobile devices accounted for 10 percent in 2014, and the percentage is increasing.

What are people doing with the mobile Internet? Some are checking their e-mail. Others are searching for information, playing games, or keeping in contact with others. People want to be connected, entertained, and informed, and they want it available when they want them.

Mobile is different

Making a website work on a mobile device is vastly different from making it work on a desktop machine. There are so many things that were features on a desktop browser that are now a hindrance to work with on a mobile device. Let's look at both the good and the bad of what makes mobile different.

The good

Even with all the negatives, mobile application development is an exciting field. There are a lot of good features available for mobile web applications. Modern smartphones offer sensors and tools that would be strange on a desktop, but are vital for mobile apps. They bring a world of information to the user, and let the user make notes and share things where they are, instead of later that day when they boot up their desktop. Let's look at these features in more detail.

Access to phone sensors

Mobile devices come with a number of sensors built in. These sensors can test the device's orientation, acceleration, and even location. Location can be collected through a built-in GPS device, cell phone signal triangulation, or based on the location of your Wi-Fi signal. Newer sensors within the phone can relay battery strength. Also, some third-party hybrid tools provide access to more phone features, such as the memory, contact lists, and the camera.

Instant access to the app

Users no longer have to write down a URL to pull up at home. Now, they can speak it into the browser, or take a picture of a QR Code to access a website. That instant access gets more people to use your application right away, cutting the risk of forgetting your app's location.

With instant access to the application, users can collect data where they are, instead of going home to input information. The user can take a picture of a broken fire hydrant, log in to a community issues app, and report the problem to the proper authorities. Field workers can collect feature data in the field and check it in. Volunteer geographical data collection can be used for citizen science, municipal issue tracking, and a host of other applications.

Instant access to the user

Unlike a desktop application that may be used by anybody sitting at a library computer kiosk, mobile phones are more likely to have a single user. Application usage can be tied in to user profiles to give you a more complete picture of how your app is used by different demographics. The possibilities are endless.

The bad

You spend hours putting together a gorgeous website that looks perfect on your monitor. You test your site on three or four different browsers, and like what you see. It looks great; it works great. It's bound to be a success.

Then, you run into a friend and want to show them your amazing website, but all you have is your smartphone. No problem, you think, as you type in site's URL into your phone's browser. What comes up is a usability nightmare. Parts of the screen are cut off. Controls don't work like you planned. You can't navigate through the menus properly. Overall, it's a mess.

Now, let's look at some of the pain points of mobile web development.

So many screen sizes

Back in the old days of the Internet, you only had to worry about a few monitor sizes. As long as you made a site that looked good on a 1024x768 monitor, you were okay. Some people had the money to afford larger monitors, while some others had smaller, but there wasn't a big difference.

Now, smartphones, tablets, and other devices have screens that range from four inches corner-to-corner to flat-screen television sizes and, because screen technology has improved so much, the smaller screens have a pixel density 1.5, 2, or even 3 times that of a standard desktop monitor. Websites that were easy to read on a desktop become squished and smaller on a smartphone.

As the number of screen resolutions increases, so does the number of tests you need to perform on your website. Does the site look as good on a three inch wide phone as on an HD television? Does it scale well?

Fingertip accuracy

Another feature lost when moving from desktop to mobile applications is the high accuracy input of a mouse. Whether you use a mouse, a laptop trackpad, or a stylus pen with your computer, you have a mouse pointer that provides fine manipulation of your content. On mobile devices, you have to account for rather large fingers that may click more than one of your cleverly sized buttons. Also, the little closeout buttons you created for your site may be too hard to close with some of the large fingerprints out there.

Along with losing the accuracy of a mouse pointer, you also lose mouse hover events. Everything from simple tooltips to CSS-powered collapsible menus no longer work as expected in a mobile browser, because there are no hover events to listen to and work with. Your old code from five to ten years ago won't work the same in the mobile web era.

Battery life

Dealing with battery life can be another hindrance to mobile development. Repeated access to location data and constantly monitoring advertising can drain the battery on mobile devices. While this information is handy to have, it comes at a price. Remember that not everybody has a full charge on their phone, and not every battery will run for hours.

More devices

We mentioned the multitude of screen sizes before, but that's just the beginning. There are a wide variety of devices out there, running Android, iOS, Windows Phone, and other operating systems. Each one has a choice of a number of web browsers, and even those web browsers may be at different version numbers. With all this, support for the latest and greatest web features can be spotty, depending on the feature. You'll have to decide what you're willing to support, and what devices you're willing to purchase for testing.

Now that we've looked at why we should be building mobile applications, let's look at the tools we have available through the ArcGIS JavaScript API.

The ArcGIS compact build

The ArcGIS JavaScript API can be loaded as a more compact library for mobile browsers. The compact build, as it is called, packs the bare minimum of the libraries needed to view map applications in a mobile browser. Other modules can be downloaded through Dojo's `require()` statements, but many will not be preloaded with the initial library.

Modules included in the compact build

The ArcGIS compact build contains all the modules necessary to build a web map application. It loads the same as the regular ArcGIS API for JavaScript, using `require()` and `define()` statements. It also comes with the most frequently used modules, such as `esri/Map` and `esri/layers/ArcGISDynamicMapServiceLayer`, to quickly load your maps while using the least bandwidth.

What's not included

With all the functionality the ArcGIS JavaScript compact build offers, you'd think they must sacrifice something. The first thing that the compact build gives up is weight. At 179.54 KB in version 3.13, the library weighs in 107.26 KB under its bulkier cousin. The regular build comes with a number of libraries preloaded, while the compact build uses `require()` or `define()` statements to request those modules separately. With this, you have better control over what library parts you send to the user.

Another item sacrificed in the ArcGIS JavaScript API compact build is the reliance on the `dijit` namespace. The first thing you'll notice is that the popups are replaced with a more simplified versions. Also, if you like the graduated zoom slider to zoom your map in and out, you can forget it in the compact build. It only supports the **+** and **−** buttons to zoom the map in and out. If you have widgets that rely on the `dijit/_Widgetbase` library, those can be downloaded separately through `require()` and `define()` statements.

What does this mean to you?

The ArcGIS JavaScript compact build provides much of the same functionality as the regular build does. There are a few differences in some of the controls, but they both present the same maps and information. The smaller library size is perfect for dropping a map into an existing application, or for using other libraries, such as Angular, Knockout, or jQuery, to handle other component interactions. If you don't have a dependence on the few features lost by using the compact build, it's worth trying.

For more information on the ArcGIS JavaScript API compact build, look at the ArcGIS JavaScript API documentation at `https://developers.arcgis.com/javascript/jshelp/inside_compactbuild.html`.

ESRI Leaflet

The `Leaflet.js` library provides another alternative to the ArcGIS JavaScript API. It's a lightweight library that can show maps on a large range of browsers. Leaflet works well with any tiled map services, and points, lines, and polygons can be added through **geojson**, a popular open-source JSON format for geographical data. The library can support different tools and data sources with plugins. There is a rich plugin ecosystem for the Leaflet library, with more tools and data source plugins developed daily.

ESRI has released the ESRI Leaflet plugin so that Leaflet maps can use ArcGIS Server Map Services. According to the ESRI Leaflet GitHub page, it supports the following map service layers:

- ESRI basemap services
- Feature services
- Tiled map services
- Dynamic map services
- ImageServer map services

 For more information about the `Leaflet.js` library, you can visit `http://leafletjs.com/`. For books on the library, you can check out *Leaflet.js Essentials* by Paul Crickard III, or *Interactive Map Designs with Leaflet JavaScript Library How-to* by Jonathan Derrough.

Dojox mobile

Don't you wish you could create an application that mimics the style of a mobile device, while looking like a native app? That's what some of the contributors to the Dojo framework thought, and that led to the modules in `dojox/mobile`. The modules provide controls that match many of the UI elements in native mobile apps, mimicking them in form and function. With the widgets in this library, buttons and sliders look like iPhone buttons and sliders on Safari, while appearing as native Android buttons and sliders on Android-based browsers.

The `dojox/mobile` modules provide a visual interactive framework that mimics native mobile apps. Unlike their `dijit` form counterparts, the `dojox/mobile` user controls do not use so many HTML elements, improving the download speed and memory usage. The UI elements work well with other `dojox/mobile` controls, from the `dojox/mobile/View` that takes up the whole screen, down to the last `dojox/mobile/Button`. Let's take a look at a few of them.

Dojox mobile views

The `dojox/mobile/View` module provides a visual separation between parts of the application. Views are full page containers that can be navigated to and from by swipes or button presses. These are somewhat analogous to the `dijit/layout/ContentPane` in how they organize content.

Related to the `dojox/mobile/View`, the `dojox/mobile/ScrollableView` provides extra scrolling functionality in a way mobile users expect. In many mobile devices, when the user swipes the screen to scroll down the page, that user expects the page will continue to scroll until it slows to a stop. `ScrollableView` implements that inertial scrolling down the page.

 As `ScrollableView` scrolling events interfere with panning a map on a touchscreen interface, you should not add an interactive map to this view. `ScrollableView` is better suited for forms and content that may extend beyond the height of the screen.

Working with touch

If you're used to working with mouse events in JavaScript, touch events can be a bit confusing. The `touchstart` and `touchend` events look equivalent to the `mousedown` and `mouseup` events.

In current versions of the ArcGIS JavaScript API, many of the touch events are already handled by the API modules. When working with the map, you don't need to assign a `map.on("touchstart")` event listener on top of a `map.on("click")` event listener. The `map.on("click")` event listener handles it for you. The same goes for any Dojo widgets and controls. That's one less thing you have to do to make your application mobile-ready.

Speaking of the map, there are touch events available that make some navigation tools obsolete. You can pinch or spread your fingers on the screen to zoom in and out respectively. Panning can be controlled by dragging your finger across the map. These actions remove the need for zoom in, zoom out, and pan buttons, which can free up valuable screen real estate.

Gestures

JavaScript handling of complicated mobile gestures hasn't been as smooth as it has been in native applications. Native applications are developed to distinguish between a tap and a tap-and-hold, for instance. By default, many JavaScript-based applications treat them both as a click.

The Dojo framework that comes with the ArcGIS JavaScript API has some experimental libraries to handle gestures, with the `dojox/gesture` modules. These modules allow you to assign events using `dojo/on`, as in the following snippet:

```
require(["dojo/on", "dojox/gesture/tap"], function (dojoOn, tap) {
  dojoOn(node, tap, function (e) {…});
});
```

For simple gesture definition, `dojox/gesture` modules allow you to define tap and swipe events using `dojox/gesture/tap` and `dojox/gesture/swipe` respectively. With tap events, you can define single tap, double tap, and tap-and-hold events. For swipe events, you can define events at the beginning and end of the swipe event. In the following, you can see a code snippet implementing it:

```
require(["dojo/on", "dojox/gesture/swipe"], function(on, swipe){
  on(node, swipe, function(e){ … });
  on(node, swipe.end, function(e){ alert("That was easy.");});
});
```

You can't find a gesture that does what you want? With the `dojox/gesture/Base` module, you can define your own custom gestures. As of now, you have to define your own methods to handle gestures such as rotating, pinching, and spreading your fingers. At some point, there will be more general support for those gestures, but not as of the time of writing.

 If you would like to learn more about handling touch and gestures in Dojo applications, you can visit `https://dojotoolkit.org/reference-guide/1.10/dojox/gesture.html`.

Our application

As our story continues, we receive a call from the city of Hollister, California, regarding their incident reporting app. They like the application, and it works great for the receptionist who takes phone calls about those issues. Now, they want a version that's more mobile-friendly, and they've come to us for help. It's time for us to take our knowledge of mobile apps and create a tool they can use from a smartphone in the field.

The original incident reporting app was built using typical `dijit/layout` elements, where every panel had a place on the screen. Now, we have to consider that there's not enough room on a smaller screen for everything. Instead, we need to organize each panel into its own separate view. We'll need to control how we navigate between these views, and use the appropriate controls that work well with a mobile device.

We'll use the ArcGIS JavaScript API compact build, along with the `dojox/mobile` modules, to create a mobile-friendly web application. We'll put the map in one view, the incident picker in a second view, and a more detailed reporting form in the third view. For all of these, we'll use `dojox/mobile` user interface components, along with the ArcGIS JavaScript API editing widgets, to create not just a mobile-friendly, but a user-friendly reporting experience as well.

Changing the layout

We will begin creating the mobile application by loading the ArcGIS compact build into the `index.html` file. In the head of the HTML document, we'll change the link to the ArcGIS JavaScript API to load the compact build. We'll keep the `esri.css` file and our own style sheet reference, but we can remove the `claro.css` style sheet reference, since our application won't need it. Our `index.html` file should look like the following:

```
<head>
  ...
  <!-- note that the claro.css stylesheet is removed -->
  <link rel="stylesheet"
    href="https://js.arcgis.com/3.13/esri/css/esri.css" />
  <link rel="stylesheet" href="./css/style.css" />
  <script type="text/javascript">
    dojoConfig = { async: true, isDebug: true };
  </script>
  <script src="https://js.arcgis.com/3.13compact/"></script>
</head>
...
```

The body of our application has three actionable parts. There's a map where we place our incidences. There's also a panel where we select what incident is on the map. Finally, there is a form where we fill out more information about the incident. We'll lay those out into different views: mapview, incidentview, and attributeview. Inside each view, we'll add the headings and controls we need for our application. It should look like the following:

```
<body>
  <div id="mapview" data-dojo-type="dojox/mobile/View">
    <h1 data-dojo-type="dojox/mobile/Heading">
      Incident Reporting App
    </h1>
    <div id="map" >
      <div id="locatebutton"></div>
    </div>
  </div>
  <div id="incidentview" data-dojo-type="dojox/mobile/View">
    <h2 data-dojo-type="dojox/mobile/Heading"
    data-dojo-props="back:'Map',moveTo:'mapview'">
      Incident
    </h2>
    <div id="pickerdiv"></div>
  </div>
  <div id="attributeview" data-dojo-type="dojox/mobile/View">
    <h2 data-dojo-type="dojox/mobile/Heading"
    data-dojo-props="back:'Incident',moveTo:'incidentview'">
      Description
    </h2>
    <div id="attributediv"></div>
  </div>
  ...
</body>
```

In the preceding code, we've added the familiar data-dojo-type attributes to create the dojox/mobile/View modules on the page. Inside each view, we have a dojox/mobile/Heading module element, to show a heading at the top of the page. The heading also doubles up as a sort of a button bar, into which we can put back buttons inside. In the data-dojo-props attribute of the headings, we define a back button with the back attribute defining the button text, and the moveTo attribute defining the view it switches to.

Modifying the JavaScript

In our app.js file, we'll need to modify the require statement in order to load the appropriate modules for the mobile library. Instead of loading the dijit/layout modules for setting up the layout, we'll need to add the dojox/mobile equivalent. In the require() statement in the app.js file, modify the code to add the following:

```
require([
  "dojox/mobile/parser",
  "dojo/dom",
  "dojo/on",
  "esri/config",
  "esri/map",
  "esri/graphic",
  "esri/layers/FeatureLayer",
  "esri/layers/ArcGISDynamicMapServiceLayer",
  "esri/symbols/SimpleMarkerSymbol",
  "esri/geometry/Extent",
  "esri/dijit/editing/Editor",
  "esri/dijit/editing/TemplatePicker",
  "esri/dijit/editing/AttributeInspector",
  "esri/dijit/LocateButton",
  "esri/tasks/query",
  "dijit/registry",
  "dojox/mobile/Button",
  "dojox/mobile",
  "dojox/mobile/deviceTheme",
  "dojox/mobile/compat",
  "dojox/mobile/View",
  "dojox/mobile/Heading",
  "dojox/mobile/ToolBarButton",
  "dojo/domReady!"
], function (
  parser, dojoDom, dojoOn,
  esriConfig, Map, Graphic,
  FeatureLayer, ArcGISDynamicMapServiceLayer,
  MarkerSymbol, Extent,
  Editor, TemplatePicker, AttributeInspector,
  LocateButton, Query, registry, Button
) {
```

As you can see, we replaced the normal dojo/parser with a dojox/mobile/ parser equivalent. We've added the dojox/mobile base class, the dojox/mobile/ deviceTheme that loads the appropriate theme based on your browser, and dojox/ mobile/compat so that the site can also be seen on older desktop browsers such as Internet Explorer. For the elements we want to see out of the dojox/mobile library, we've loaded the View, the Heading to view title data, and the ToolBarButton to add buttons to the heading.

Working with the map on mobile

Let's focus on making the map work. In our mobile application, we've added the map within the mapview div. We've set the width and height of the map to 100% in our style.css file. The map should load as normal, right?

When we load the map as it is, especially from a larger browser, we find that the map doesn't stretch all the way to the bottom. Using our favorite tools for examining DOM elements, we find that height of the map div has been set inline to 400px. Where have we seen this before?

After examining the DOM elements of features around the map, we see that the view's height hasn't been set. By default, the mapview div's height depends on the height of its content. As its height has not been defined, the map sets its height to 400px. To fix this, we need to manually define the mapview div's height in our style.css file. We'll also stop the mapview div from scrolling by setting its overflow-y to hidden. This will remove any unsightly scrollbars on our map, which may interfere with map navigation:

```
#mapview {
  height: 100%;
  overflow-y: hidden;
}

#map {
  width: 100%;
  height: 100%;
}
```

Fixing the LocateButton

The LocateButton, which uses the browser's GPS functionality to center the map on our location, has moved around. It appears that addition of the dojox/mobile/ Heading and the mobile buttons to zoom in and out have caused our LocateButton to be displaced. We can use our favorite browser DOM explorer to reposition the Locate button to a good spot, and then include it in the style.css file to make it more permanent. The style for the LocateButton should look something like the following:

```css
.LocateButton {
  position: absolute;
  left: 21px;
  top: 130px;
  z-index: 500;
}
```

When you're done, you should have a map that looks like the following:

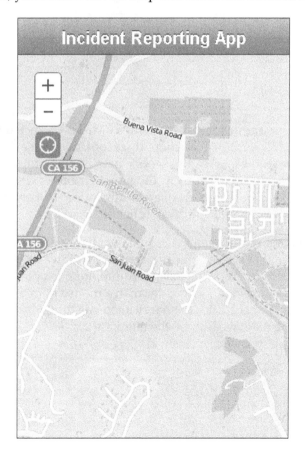

Working with the editor widgets

Now that we're using the ArcGIS compact build, we don't have access to the popup dijit that we used with the attachment editor. We don't have many of the other dijit-based modules either. This application may require a bit more work to make it mobile-ready.

Template picker

To select from the list of incidents, we chose the `esri/dijit/editing/TemplatePicker` module to create buttons to select an incident. Now, we'll keep using it, but we're looking at it in a different view. The original one presented a vertical list of buttons down the side of the page for presenting incidents. Now, we're going to remove those settings and define the template picker more normally. The code for initializing the template picker should look like the following:

```
function startEditing () {
  ...
  picker = new TemplatePicker({
    featureLayers: [ layer ],
    style: "width:100%;height:auto;",
    grouping: false
  }, "pickerdiv");
  picker.startup();
  ...
```

To access the picker for our incidents, or to edit a currently selected feature, we need to call the `showInspector()` function. If we look over the existing function, it attempts to select features in a feature service based on a point around where we click. It uses the map's `infoWindow` to show the attribute editor. Now that we're using another location to edit the feature attributes, we need to modify the `showInspector()` code to handle our new functionality.

Our first step in making our `showInspector()` function work on a mobile is to tweak the surface area that is selected compared to that if we click on the map. Currently, it creates a two pixel wide extent around our click point. We can expand it to 10 pixels, because our fingers are wider than a mouse pointer. Also, we need to modify the `callback` function after the query succeeds. If there are no features in the location clicked on the map, we'll show the template picker. If none is selected, we'll tell it to go to the attribute inspector instead:

```
function showInspector(evt) {
  var selectQuery = new Query(),
  point = evt.mapPoint,
  mapScale = map.getScale();
  selectQuery.geometry = new Extent({
```

```
        xmin: point.x - mapScale * 5 / 96,
        xmax: point.x + mapScale * 5 / 96,
        ymin: point.y - mapScale * 5 / 96,
        ymax: point.y + mapScale * 5 / 96,
        spatialReference: map.spatialReference
      });

      incidentLayer.selectFeatures(selectQuery,
      FeatureLayer.SELECTION_NEW, function (features) {
        if (!features.length) {
          goToPicker(point);
        } else {
          goToAttributeInspector('mapview', features[0]);
        }
      });
      }
```

In our `goToPicker()` function, we'll switch from `mapview` to `incidentview`. We'll do this by using the `performTransition()` method provided by `dojox/mobile/ View`. It accepts up to five arguments: an `id` for another view to see, a number-based direction (either `1` or `-1`), a transition style, and an object that defines `this` for the `callback` function that rounds out the fifth argument. We'll tell `mapview` to make the transition to `incidentview`, moving from the right with a `slide` animation, and we'll add a `callback` function once the process finishes:

```
    function goToPicker(point) {
        registry.byId('mapview').performTransition('incidentview', 1,
        'slide', null, function() {
    ...
      });
    }
```

When we try to run the `goToPicker` function as it is, it goes to `incidentview`, but we don't see any incidents. This is due to an interesting feature of some dojo widgets. When not visible, the widgets set their widths and heights to `0`, effectively becoming invisible. We need to refresh the `TemplatePicker` internal grid and clear the template selection.

When the user selects a feature from the list, we need something to convey our selection to the attribute inspector. We'll also add a single fire event using the `once()` method in the `dojo/on` module, and attach it to the `selection-change` event of the `TemplatePicker` widget. From there, we'll collect the selected attributes, the current date, and a few other attributes, and pass them to a function that adds the incident to the map. The function should look like the following:

```
    registry.byId('mapview').performTransition('incidentview', 1,
    'slide', null, function() {
```

```
//refresh the grid used by the templatePicker.
picker.grid.render();
picker.clearSelection();
// on picker selection change…
dojoOn.once(picker, 'selection-change', function () {

var selected = picker.getSelected();
 if (selected) {
  // log the date and time
  var currentDate = new Date();
   var incidentAttributes = {
     req_type: selected.template.name,
     req_date:(currentDate.getMonth() + 1) + "/" +
     currentDate.getDate() + "/" + currentDate.getFullYear(),
     req_time: currentDate.toLocaleTimeString(),
     address: "",
     district: "",
      status: 1
    };
   addIncident(point, selected.symbol, incidentAttributes);
    }
   });
  });
 }
```

For the `addIncident()` function, we'll add our point location, symbol, and attributes into a graphic. From there, we'll add the graphic feature to the editable `incidentLayer`. Once we have completed that, we'll attempt to select it again using the `incidentLayer` layer's `selectFeatures()` method, then send the result to the `goToAttributeInspector()` function. We'll pass along the name of the current view (`incidentview`), as well as the first feature that's selected. It should look like the following:

```
function addIncident(point, symbol, attributes) {
   var incident = new Graphic(point, symbol, attributes);
// add incident to the map so it can be selected and passed to the
// attribute inspector.
incidentLayer.applyEdits([incident],null,null).then(function () {
  var selectQuery = new Query(),
        mapScale = map.getScale();
  selectQuery.geometry = new Extent({
    xmin: point.x - mapScale / 96,
    xmax: point.x + mapScale / 96,
    ymin: point.y - mapScale / 96,
    ymax: point.y + mapScale / 96,
```

```
      spatialReference: map.spatialReference
    });

    // must select features before going to attributeInspector
    incidentLayer.selectFeatures(selectQuery,
    FeatureLayer.SELECTION_NEW, function (features) {
      if (features.length) {
        // fill in the items
        goToAttributeInspector(features[0], "incidentview");
      }
    });
    });
  }
```

If all works correctly, you should be able to access your incident picker with incidentview, and it should look like the following:

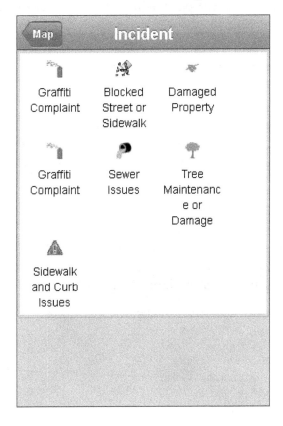

Attribute inspector

Now that the map and the template picker have been properly updated for the mobile application, it's time to look at the third stage of our incident reporting application. Using our previous code, we'll select the incident through the `FeatureLayer` selection. We'll then load the attribute inspector and edit the data. Finally, we'll save the feature data, including any images, to the feature service.

In the previous desktop application, we loaded the attribute inspector in the map popup. However, we don't have the same popup widget we had before in the ArcGIS compact build. This one won't be so editable on the map screen. However, we have plenty of screen real estate on the attribute view, so we'll load the inspector there. Also, note that we'll remove any of the events where the map popup loads or changes the attribute inspector.

First, we need to create a location on the page for the attribute inspector within the attribute view. In the `index.html` page, within the `attributeview` div element, we'll add a div element with an `id` of `attinspector`. When our application loads, it will create an attribute inspector in this location. It should look like the following:

```
<div id="attributeview" data-dojo-type="dojox/mobile/View">
  <div id="attinspector"></div>
</div>
```

In our `app.js` file, we will still use the `generateAttributeInspector()` function that is called by the `startEditing()` function when the map is loaded. However, the `generateAttributeInspector()` function will need a few changes to work with its more permanent surroundings. We will need to do the following:

- Initialize and start up the attribute inspector where the `attinspector` div element is located

- Remove any references to the `infoWindow` property of the map

- When the changes are applied to the `generateAttributeInspector()` function, it should look something like the following:

```
function generateAttributeInspector(layer) {

    layerInfos = [{
      featureLayer: layer,
      showAttachments: true,
      isEditable: true,
      showDeleteButton: false,
      fieldInfos: [
        {'fieldName': 'req_type', 'isEditable':true,
         'tooltip': 'What\'s wrong?', 'label':'Status:'},
```

```
          {'fieldName': 'req_date', 'isEditable':false,
          'tooltip': 'Date incident was reported.',
          'label':'Date:'},
          {'fieldName': 'req_time',
          'isEditable':false,'label':'Time:'},
          {'fieldName': 'address', 'isEditable':true,
          'label':'Address:'},
          {'fieldName': 'district', 'isEditable':true,
          'label':'District:'},
          {'fieldName': 'status', 'isEditable':false,
          'label':'Status:'}
      ]
  }];
//"","req_date","req_time","address","district","status"

  attInspector = new AttributeInspector({
    layerInfos: layerInfos
  }, "attinspector");

  attInspector.startup();

  //add a save button next to the delete button
  var saveButton = new Button({ label: "Save", "class":
  "saveButton"});
  domConstruct.place(saveButton.domNode,
  attInspector.deleteBtn.domNode, "after");

  saveButton.on("click", function(){
    updateFeature.getLayer()
      .applyEdits(null, [updateFeature], null);
  });

  attInspector.on("attribute-change", function(evt) {
    //store the updates to apply when the save button is
    clicked
    updateFeature.attributes[evt.fieldName] =
    evt.fieldValue;
  });

  attInspector.on("next", function(evt) {
    updateFeature = evt.feature;
    console.log("Next " +
    updateFeature.attributes.objectid);
  });
```

```
        attInspector.on("delete", function(evt){
          evt.feature.getLayer()
            .applyEdits(null,null,[updateFeature]);
          map.infoWindow.hide();
        });
        // content after this was deleted.
      }
```

Once we have made the changes, we can run the application in a browser and check out the attribute inspector. After clicking on a troublesome location on our map, and identifying the incident with `TemplatePicker`, we should be able to view and edit the incident attributes with the attribute inspector.

Trouble in the app

Oops, we've run into a bit of a problem with the application. We tested the app by tapping on a spot on the map to report an incident. We selected the incident type from the template picker, and it made the selection. After a couple of seconds, it switched over to the attribute inspector, and we got the following:

The attribute inspector form is very unsightly, and doesn't behave in the way it did as a desktop web application. The user controls to edit the feature attributes don't work very well. How could this happen?

The issues with the attribute inspector actually lead back to something we did at the beginning of this application. We removed the `claro.css` file at the head of the web page and, along with it, removed any other `dijit` references. This act saved significant bandwidth on our application, but we lost the styling and functionality of the user controls in the attribute inspector. Now, it's not going to do what we want it to.

Rebuilding the attribute inspector

There is another way, however. We can create our own form for updating the feature attributes. We can use form elements from the `dojox/mobile` modules to make our own form, instead of using the attribute inspector. Also, on closer inspection, the attachment editor portion of the attribute inspector worked well. We can load the attachment editor after our custom form, and use it to save images to the features.

Creating the form

To make the custom form, we're going to need to load a few `dojox/mobile` modules to parse. In the `require()` list in our `app.js` file, we'll add the `dojox/mobile/RoundRect` module to create a rounded body for the form. We'll also use `dojox/mobile/TextBox` for text entry, as well as the combination of `dijit/form/DataList` and `dojox/mobile/ComboBox` to create a mobile drop-down menu. We're also going to use `dojox/mobile/Button` to save our changes. Our `require` statement should look like the following:

```
require([…, "dojox/mobile/RoundRect", "dojox/mobile/TextBox",
  "dijit/form/DataList", "dojox/mobile/ComboBox",
   "dojox/mobile/Button", "dojo/domReady!"], function ( … ) {
…
});
```

Next, we'll modify the attribute view to make the form for editing the incident attributes. We'll use `DataList` and `ComboBox` in `index.html` as a selection tool for the incident type. In this way, if the wrong type is selected, the user will be able to correct it. Next, we'll use `Textbox` to record the `Address` and `District` attributes. The date, time, and status are read-only at this point, since we don't want the reporter to change the date and time of the incident and whether the incident has been opened or closed. Finally, we'll add a `Save` button to the form to save the results. Once we add these to `index.html`, the file should look like the following:

```
<div id="attributeview" data-dojo-type="dojox/mobile/View">
  <h2 data-dojo-type="dojox/mobile/Heading"
    data-dojo-props="back:'Incident',moveTo:'incidentview'">
      Description
  </h2>
  <div data-dojo-type="dojox/mobile/RoundRect">
    <label>Request Type:</label>
    <select data-dojo-type="dijit/form/DataList"
      data-dojo-props='id:"incidentDataList"'>
        <option>Graffiti Complaint</option>
        <option>Blocked Street or Sidewalk</option>
        <option>Damaged Property</option>
        <option>Sewer Issues</option>
        <option>Tree Maintenance or Damage</option>
        <option>Sidewalk and Curb Issues</option>
    </select>
    <input type="text" data-dojo-type="dojox/mobile/ComboBox"
      data-dojo-props='list:"incidentDataList",
        id:"incidentSelector"' />
    <br />
    <label>Date:</label>
    <span id="incidentDate"></span>
    <br />
    <label>Time:</label>
    <span id="incidentTime"></span>
    <br />
    <label>Address:</label>
    <input type="text" data-dojo-type="dojox/mobile/TextBox"
      data-dojo-props='id:"incidentAddress"' />
    <br />
    <label>District:</label>
    <input type="text" data-dojo-type="dojox/mobile/TextBox"
      data-dojo-props='id:"incidentDistrict"' />
    <br />
```

```
<label>Status: </label>
<span id="incidentStatus"></span>
<br />
<button data-dojo-type="dojox/mobile/Button"
  data-dojo-props="id:'saveButton'">Save</button>
...
</div>
```

Finally, we need to modify our JavaScript to handle the form input and output. We'll create two functions, `setupIncident()` and `saveEdits()`, to load and save the data from the incident details form. The `setupIncident()` function will accept the feature to be modified as an argument. Also, since `setupIncident()` can be called when tapping on an existing incident on the map, or after selecting an incident type in `TemplatePicker`, we'll pass the view name along with the feature data so it can move to the incident details view:

```
function setupIncident(feature, view) {
  var attr = feature.attributes;
  updateFeature = feature;
  registry.byId("incidentSelector").set("value", attr.req_type);
  dojoDom.byId("incidentDate").innerHTML = attr.req_date;
  dojoDom.byId("incidentTime").innerHTML = attr.req_time;
  registry.byId("incidentAddress").set("value", attr.address);
  registry.byId("incidentDistrict").set("value", attr.district);
  dojoDom.byId("incidentStatus").innerHTML = attr.status;

  attInspector.showAttachments(feature, incidentLayer);
  registry.byId(view).performTransition('attributeview', 1,
  'slide');
}
```

The `saveEdits()` function will collect the values from the form, add those values as feature attributes, and save the feature back to the geodatabase:

```
function setupIncident() {
  ...
}
function saveEdits(){
  // add updated values
  updateFeature.attributes.req_type =
  registry.byId("incidentSelector").get("value");
  updateFeature.attributes.address =
  registry.byId("incidentAddress").get("value");
  updateFeature.attributes.district =
  registry.byId("incidentDistrict").get("value");
```

```
      // update the feature layer
      updateFeature.getLayer().applyEdits(null, [updateFeature],
      null);
      // move back to the map view
      registry.byId("attributeview").performTransition("mapview",
      -1, 'slide');
    }
```

The attachment editor

The last feature we will be implementing is the addition of photos to the incident reports. The previous version used the attachment editor that was part of the attribute inspector. Now that we're implementing our own entry form, we need to include the attachment editor separately.

The first step in adding the attachment editor in our application is to add the module reference in the app.js file require statement. According to the API documentation, the module to use is in esri/dijit/editing/AttachmentEditor. We'll add the reference in the following code:

```
require([…
  "esri/dijit/editing/TemplatePicker",
  "esri/dijit/editing/AttachmentEditor",

  …
], function (…
  TemplatePicker, AttachmentEditor,

  …
) {  });
```

We'll initialize the attachment editor after the TemplatePicker in our startEditing() function and assign it to the variable attachmentEditor. We need the scope of the attachmentEditor to fit within the whole application since we'll be connecting it with feature data in other functions. You can see the additions highlighted in the following code:

```
require([…
  "esri/dijit/editing/TemplatePicker",
  "esri/dijit/editing/AttachmentEditor",

  …
], function (
  TemplatePicker, AttachmentEditor,

  …
) {
```

```
var attachmentEditor;
...
function startEditing () {
  // add the Locate button
  var locator = new LocateButton({map: map}, "locatebutton");
  var incidentLayer = map.getLayer("incidentLayer");

  picker = new TemplatePicker({
    featureLayers: [ layer ],
    style: "width:100%;height:auto;",
    grouping: false
  }, "pickerdiv");
  picker.startup();

  attachmentEditor = new AttachmentEditor({}, "attributediv");
  attachmentEditor.startup();
  ...
}
...
});
```

When we connect our editing form with our data in the `setupIncident()` function, we also need to connect the `attachmentEditor` to its data. After we've updated the editing form with the feature values, we'll call the `showAttachments()` method of the `attachmentEditor`. This method accepts the feature, as well as the layer to be edited. The attachment editor will handle how to display existing attachments, and how to add new ones. The code changes should look like the following:

```
function setupIncident(feature, view) {
  var attr = feature.attributes;
  updateFeature = feature;
  registry.byId("incidentSelector").set("value", attr.req_type);
  dojoDom.byId("incidentDate").innerHTML = attr.req_date;
  dojoDom.byId("incidentTime").innerHTML = attr.req_time;
  registry.byId("incidentAddress").set("value", attr.address);
  registry.byId("incidentDistrict").set("value", attr.district);
  dojoDom.byId("incidentStatus").innerHTML = attr.status;

  attachmentEditor.showAttachments(feature, incidentLayer);
  registry.byId(view).performTransition('attributeview', 1,
    'slide');
}
```

Finally, we need to supply an element to `index.html`, in which we will attach the attachment editor widget. At the bottom of the `attributeview` element, underneath the editing form, we'll add a `div` element with the `id` of `attributediv`. That portion of our `index.html` page should look like the following:

```
<div id="attributeview" data-dojo-type="dojox/mobile/View">
  <h2 data-dojo-type="dojox/mobile/Heading"
    data-dojo-props="back:'Incident',moveTo:'incidentview'">
  Description
  </h2>
  <div data-dojo-type="dojox/mobile/RoundRect">
    ...
    <button data-dojo-type="dojox/mobile/Button"
    data-dojo-props="id:'saveButton'">Save</button>
  </div>
  <div id="attributediv"></div>
</div>
```

When you run the application and begin reporting an incident, you should eventually see a form that looks like the following:

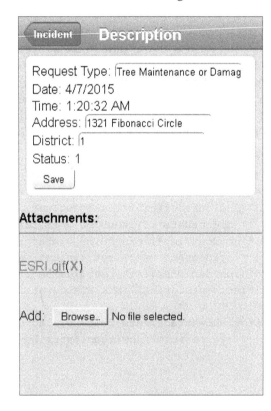

The end result

After modifying our application layout and behavior, we now have a useful mobile reporting tool. A citizen could load this page in their browser and report any problems they find. City workers could also use this to report incidents. Your application should look like this:

More work could be done to improve the application. Currently, it requires a constant Internet connection to make edits. If your mobile device is in an area with bad network coverage, any edits will be lost.

Summary

In this chapter, we have examined what makes a website mobile. We looked at the requirements for a mobile application, in screen real estate, functionality, and bandwidth. We looked at the ArcGIS compact build to use on mobile applications. We also modified an existing application previously formatted for desktop use, and made it mobile-ready.

In the next chapter, we'll investigate how to write testable code using test-driven development.

10
Testing

When you publish a website or a web application to the public, users expect it to work. Links are supposed to take their places, and buttons are supposed to do things. They may not have the technical knowhow to look in good their browser console or observe network traffic. They may even have good reason to disable JavaScript on their browser.

The truth is, sometimes websites don't work because of hardware, software, or user error, but at other times the cause comes from the developer's buggy code. Maybe the code was pieced together from a cut-and-paste collage of code samples and stack overflow answers. Then the developer rushes to test the application once or twice, using the same workflow they use every time they write a component.

Maybe, though, something changed outside of the developer's control. Maybe a feature attribute that once held numeric values now contains strings. Maybe the API was updated and there was a fundamental change in the application. Maybe a REST data service is down, or maybe some bored kid is trying to input random data or malicious SQL injection code to bring down the website.

Testing through code

These are all good reasons to create tests for your applications. Testing is one method of ensuring that your code does as expected. It defines clear use cases, with clearly defined, expected results. Testing also acts as a sort of project documentation. It tells other developers who work with the project how the code is supposed to behave.

Now that you understand why testing is important, let's look at the testing methodologies used by other developers over the years.

Unit testing

Unit testing is when individual components or modules of code are loaded into a testing framework and tested individually. Tests are developed and results are compared to expected values. For object-oriented programming, individual objects can be loaded and tested without having to wait for other components to load.

JavaScript objects and modules lend themselves well to unit testing. Developers can write scripts to load modules individually, pass values into functions and objects, and see what comes out. Passing the tests ensures the JavaScript modules and objects will perform as far as expected test cases go.

However, unit testing doesn't prevent all errors. While they tests the input and output of objects and functions, they don't test how these objects interact. What happens when two or more memory-intense modules are running in the same browser? What happens to module A when input expected from module B times out? Also, unit testing doesn't address user experience concerns. It's hard to write a unit test for a map that pans slower than I think it should. Clearly, unit testing can't be the only tool in your testing tool belt.

Functional testing

Functional testing handles the testing of a single process, instead of a single unit. It makes many of the changes expected in a workflow, and checks that the results line up with what is expected. While one or more unit tests may cover the components tested in the functional test, the functional test connects them together to simulate a real-world example.

An example might be the calculation of **Value Added Tax (VAT)** on an online purchase in European countries. You might have unit tests that cover the following:

- Whether Italy has the correct VAT percentage
- What the subtotal of a list of purchased items costs
- What the total cost of a purchase would be after VAT is applied

With a functional test, you can string them all together. You could ask, given that a person in Italy purchases these items, does the total cost after VAT matches what is expected.

End-to-end testing

If unit testing examines the individual components of an application and functional testing tests a single workflow, then end-to-end testing examines the whole thing from beginning to end. The entire application is loaded into a platform specially designed for handling tests. A user script is loaded with predefined actions, such as text to enter or buttons to click. Any errors are logged by the system for review. Some testing frameworks even record screenshots or video of the testing process for review.

End-to-end testing is meant to catch the errors generated by components interacting with one another. They also tend to catch improper implementation of the components (that is human error). However, end-to-end testing can be expensive in terms of computer resources and time. The more use cases you test using end-to-end testing, the longer it will take to get results. End-to-end testing is typically saved for code releases or nightly build processes, while unit testing is expected to cover the minor code changes between releases.

There are a number of paid services available on the web for end-to-end testing. For some frameworks, such as Angular, there are testing frameworks such as **Protractor**.js that can be used for end-to-end testing, but if you want end-to-end testing that is both free and framework agnostic, you have options. Open source libraries such as Phantom.JS, Casper.JS, and web browser automation tools such as Selenium, can help you test your application from start to finish, with a bit of setup.

Testing as you code

Many developers have found that the best time to test their code is when they are writing it. It's a horrible feeling when you spend all day writing a module, only to have it fail and not know why. By creating unit tests in the middle of development, the developer receives quick feedback regarding whether part of their code succeeds or fails. Now, let's look at some ways developers write tests as they code.

Test-Driven Development

Test-Driven Development (TDD) is the process of writing unit tests and end-to-end tests, then writing code to pass those tests. Whenever code is refactored, it runs through the same tests to ensure results come out as expected.

Developers using TDD tend to code using the following steps:

1. Write a test based on requirements.
2. Run all tests and see that the new test fails. This eliminates unnecessary tests and tests the overall testing framework.
3. Write some code to pass the test.
4. Run the tests again.
5. Refactor as necessary to clean up ugly code, names, and functionality.
6. Repeat steps 1 to 5 until done.

By writing the test before writing code, the developer keeps their focus on the module's objectives and does not write unnecessary code. The tests also give the developer confidence in the code, and lead to faster code development.

Behavior-Driven Development

Behavior-Driven Development (BDD) is a form of TDD that answers the question "how much should I test?". In BDD, tests for code modules and objects are written with descriptions that start with "It should do _____". Code tests are then written to test for those features and those only.

BDD tests help to both define acceptance criteria and document code. By looking through the test descriptions, other developers will have a better idea of what the code is supposed to do. By describing acceptance criteria up front, the developer doesn't have to write extra code for use cases the code will never experience.

Test statements

At the heart of every TDD and BDD setup is the test statement. The test statements are lines of code written in a readable format to test properties and values of results. Test statements define a simple test within unit tests or functional tests.

Depending on the library you choose for generating test statements, they may follow a BDD `should` pattern, a TDD `expect` pattern, or a TDD `assert` pattern. Some libraries may let you use one or more of these patterns for your tests. For example, using the `Chai.js` library (which we'll discuss a little later in the chapter), we can look at examples using any of the three patterns. Tests using the `should` pattern might look like the following:

```
chai.should();

country.should.be.a('string');
```

```
country.should.equal('Greece');
country.should.have.length(6);
texas.should.have.property('counties')
  .with.length(253);
```

The `expect` pattern may look something like the following:

```
var expect = chai.expect;

expect(country).to.be.a('string');
expect(country).to.equal('Greece');
expect(country).to.have.length(6);
expect(texas).to.have.property('counties')
  .with.length(253);
```

Finally, the `assert` pattern should look something like the following:

```
var assert = chai.assert;

assert.typeOf(country, 'string');
assert.equal(country, 'Greece');
assert.lengthOf(country, 6)
assert.property(texas, 'counties');
assert.lengthOf(texas.counties, 253);
```

All three formats test the same things. Their only difference is the syntax they use to get the same results. Depending on which testing formats your team likes to write, many of the testing libraries will have something in that flavor.

Now that we've reviewed the general concepts behind software testing, let's look at the tools we have to work with. In the last few years, the JavaScript community has produced a number of testing frameworks. Some are specific to a framework, such as Protractor for Angular, while others work with almost any library or framework, such as Mocha, Jasmine, or Intern. In this chapter, we're going to examine two of them: Intern and Jasmine.

Intern testing framework

Intern (`https://theintern.github.io/`) is a testing framework for testing websites and applications. Its website boasts that not only can it test plain JavaScript applications, but also test server-side websites built with Ruby and PHP, as well as mobile iOS, Android, and Firefox apps. Intern supports AMD and promises asynchronous testing.

If you're using Node.js in your development environment, Intern may integrate well. If you use Grunt, Intern comes with its own Grunt tasks for easy integration into your existing workflow. Intern also works with continuous integration services such as Jenkins and Travis CI.

Intern works in most modern browsers, including Android Browser, Chrome, Firefox, Safari, Opera, and Internet Explorer. For Internet Explorer, Intern works with versions 9 and later. If you need to test older versions of Internet Explorer, there is an `intern-geezer` module available through `npm`. The tests will work with Internet Explorer versions 6-8, which do not support EcmaScript 5, which is required by regular Intern.

Setting up your testing environment

If you're using Node.js, installation can be as simple as `npm install intern`. According to the website, Intern has a recommended folder structure for your projects, as shown in the following screenshot:

```
project_root/
    dist/           - (optional) Built code; mirrors the `src` directory
    node_modules/ - Node.js dependencies, including Intern
      intern/
    src/            - Front-end source code (+ browser dependencies)
      app/          - Your application code
      index.html    - Your application entry point
    tests/          - Intern tests
      functional/ - Functional tests
      support/      - Test support files
                      (custom interfaces, reporters, mocks, etc.)
      unit/         - Unit tests
      intern.js     - Intern configuration
```

The makers of Intern recommend that you keep your source code and distributable code in folders separate from the tests. In that way, the tests aren't accidently published along with the code. You may not want people finding your tests, which could expose your secured services, private API keys, and sensitive passwords.

Special requirements for the ArcGIS JavaScript API

According to reports from several users, using Intern with ESRI's hosted link to the ArcGIS API for JavaScript is not recommended. It can cause tests to fail where they shouldn't because a module and its dependencies load too slowly. The recommended solution is to download the API and the Dojo framework into folders alongside your project files. You can use Grunt, a Node.js task runner.

Writing tests for Intern

Writing tests for Intern was achieved with modules like Dojo's AMD style in mind. Once module paths have been assigned, these modules can be loaded into tests and run. Let's look at how to do that.

Unit testing with Intern

Writing unit tests for Intern is very similar to writing `Dojo` modules, with a few exceptions. Each unit test is made up of a `define()` function. The `define()` function may take either a function with a single `require` argument, or a list of string references to modules, and a function to run when all those modules have loaded. It is a common practice in Node.js modules to load them individually through a single `require()` function.

You will need to load a testing interface and an assertion library in the unit testing function. The testing interface provides a way to register testing suites. When calling the interface as a function, you provide a JavaScript object with descriptive key strings and values containing functions to run tests or objects to classify subtests. An example unit test is shown in the following snippet:

```
define(function (require) {
  var registerSuite = require('intern!object');
  var assert = require('intern/chai!assert');

  registerSuite({
    'One plus one equals two': function () {
      Var result = 1 + 1;
      assert.equal(result, 2, "1 + 1 should add up to 2");

    },
    'Two plus two equals four': function () {
      Var result = 2 + 2;
      assert.equal(result, 4, "2 + 2 should add up to 4");
    }
  });
});
```

In the preceding code, the `registerSuite` variable accepts Intern's object interface for creating tests, while `assert` accepts the `chai.assert` library loaded with Intern. The `registerSuite()` function is called, passing in two tests in an object. The first test looks at whether adding one and one will equal two, while the second test looks at adding two two's to see if that equals four. The `assert.equal()` function tests whether the result variable matches the expected result, and will throw an error with a text message should the test fail.

Test lifecycle with Intern

There may be something you need to do before you run your test in a testing suite, or before and after each test. This is the test lifecycle. Intern provides keywords in your test definitions to define what should happen during the test lifecycle. All these lifecycle functions are optional, but they will help you to create useable tests. Here is an outline of the general lifecycle of a test:

- `setup`: This function runs before any of the tests run, presumably to set up something to test against. For example, you may create a map here before you test your widget against it.

- During each test, the following events occur:

  - `beforeEach`: This function will run before each test. As an example, you might load a fresh copy of your widget.

  - The test runs.

  - `afterEach`: This function will run after each test. Here, you might destroy the widget you created in the `beforeEach` function, or you might reset variable values that may have changed during tests.

- `teardown`: This function runs after all the tests within the suite run. Here, you might destroy any maps or objects you created in the `setup` phase.

You can see an example of a unit test using the lifecycle in the following code. This one simulates loading a map and adding a widget to the map:

```
define(["intern!object", "intern/chai!expect", "esri/map",
  "app/widget"],
function (registerSuite, expect, Map, Widget) {
  var map, widget;
  registerSuite({
    setup: function () {
      map = new Map("mapdiv", {});
    },
    beforeEach: function () {
      widget = new Widget({map: map}, "widget");
```

```
    },
    afterEach: function () {
      widget.destroy();
    },
    teardown: function () {
      map.destroy();
    },
    'My first test': function () {
      // test content goes here.
    }
  });
});
```

In the preceding code, the `define` statement looks a lot like the one we're familiar with, loading multiple objects at once. Note that the `map` is created on `setup` and destroyed on `teardown`, while the `widget` is newly created before each test and destroyed after each test.

Jasmine testing framework

If you're looking for a simple BDD framework to test your application, Jasmine might have what you need. Jasmine is a framework-independent BDD library that can be used to run tests on JavaScript applications. It can either be installed through Node.js, or the library can be downloaded and copied to the test project folder.

Jasmine loads its test in an HTML file called a `SpecRunner` (short for specifications runner). This page loads the main Jasmine library, as well as all the user libraries and unit tests. Once loaded, the Jasmine library runs tests in the browser and displays the results.

Writing tests for Jasmine

Once we have the `SpecRunner` in place, it's time to write tests. Tests are written using normal JavaScript logic and a few testing objects and methods provided by Jasmine. Let's look at some of the different parts.

Writing suites, specs, and tests

Jasmine tests are organized by suites, specs, and tests. Suites make up the top level, and describe the unit tested. In fact, Jasmine tests write the suites using the `describe()` function, which describes the feature to be tested. The arguments for the `describe()` function include a string describing the feature, and a function to run the tests.

Specs for the features can be written using the `it()` function in the `describe` suite. The `it()` function, like the `describe()` function, contains a string to describe the test and a function where a test is performed on the behavior. Any errors that occur in the `it()` function lead to a failing test, while an `it()` function that runs successfully shows a passing test for the behavior. In the following you can see an example of a `describe()` and `it()` statement:

```
describe("A widget listing the attributes of the U.N.",
    function () {
      it("has its headquarters located in New York City.",
        function () {
          //tests go here
        }
      );
    }
);
```

Inside each spec is one or more tests. In Jasmine, tests are written in the expect format. Based on the object passed to the `expect` object, you have numerous tests that can be performed on it, such as whether or not it equals a specific value, or whether it comes back as a "truthy" JavaScript statement:

```
describe("A widget listing the attributes of the U.N.",
    function () {
    // unitedNations.countries is a string array of country names.
    it("contains more than 10 countries.", function () {
      expect(unitedNations.countries.length).toBeGreaterThan(10);
    });
    it("should contain France as a member country.", function () {
      expect(unitedNations.countries).toContain("France");
    });
});
```

Setup and teardown

There are times when you will need to load modules or create JavaScript objects to test. You may also want to reset them between tests and tear them down to save memory. To set up and tear down items to be tested, you would use the `beforeEach` and `afterEach` functions. Each one lets you set up and tear down values between tests, so that each test has fresh values that are repeatable. These are loaded in the suites, and are supposed to be called before any specs that require them are called:

```
describe("A widget listing the attributes of the U.N.",
    function () {
      var unitedNations;

      beforeEach(function () {
```

```
      unitedNations = new UnitedNations();
    });

    afterEach(function () {
      unitedNations.destroy();
    });

    it("should do something important", function () {
      //…
    });
  }
);
```

Ignoring tests

What if your module has a depreciated feature? What if you expect your module
to have a feature one day, but you don't want to write a test that fails now. You
could delete or comment out the code, but in doing so you lose some of the history
of your module in some respect. Jasmine has a way of disabling suites and marking
specs as pending without losing code. By replacing your describe statements with
xdescribe, and your it statements with xit, you can write tests that won't fail now,
but will be marked as "pending". You can find an example in the following code that
uses xit:

```
xit("Elbownia will join the United Nations next year.",
  function () {
    expect(unitedNations.countries).toContain("Elbownia");
  }
);
```

Red-light green-light

Many developers who write TDD and BDD tests practice red-green testing, which
is supported in Jasmine. Red stands for a failed test, while green stands for a passed
test. In red-green testing, a developer does the following:

1. Writes a necessary unit test that they know will fail (red).

2. Writes code for the object or module so that the test passes (green).

3. Refactors the code as needed, making sure it still passes the test.

4. Repeats steps 1 to 3 until both the code and tests are satisfactory.

Using the red-green testing method, the developer keeps the tests in mind as they develop module functionality. Passing tests gives the developer confidence in their code, and allows for faster development.

In Jasmine, test results on the SpecRunner HTML page are displayed using red and green colors. Red colors are shown when a test fails, while green colors show when all tests pass. The SpecRunner page will display which specification failed (if any), and the error thrown when the failure occurred. Armed with that information, the developer can fix the problem and move on to the next step in development.

Jasmine and the ArcGIS JavaScript API

Setting up your SpecRunner for applications using the ArcGIS JavaScript API is a little more challenging. The AMD nature of the Dojo framework makes loading the modules and tests a little more challenging, but it can be done.

The first step is to load the Jasmine and ArcGIS JavaScript APIs into the application. Due to all the parts that Jasmine has to handle outside our application, and because AMD modules tend to not show up when errors occur, we need to load Jasmine before the ArcGIS JavaScript API:

```html
<!DOCTYPE html>
<html>
<head>
  <meta http-equiv="Content-Type" content="text/html;
  charset=UTF-8" />
  <title>Jasmine Spec Runner</title>

  <link rel="stylesheet"
  href="https://js.arcgis.com/3.13/esri/css/esri.css" />
  <link rel="stylesheet" type="text/css"
  href="tests/jasmine/jasmine.css" />

  <script type="text/javascript" src="tests/jasmine/jasmine.js">
  </script>
  <script type="text/javascript" src="tests/jasmine/jasmine-
  html.js" ></script>

  <script type="text/javascript">var dojoConfig={};</script>
  <script type="text/javascript" src="
  https://js.arcgis.com/3.13/" ></script>
</head>
```

You will be able to load the locations of both your modules, and the tests on those modules, in the `dojoConfig` packages. We will be using the AMD loader to load the test suites as modules. In the following example, the AMD modules are in the `js` subfolder, while the spec tests are in the `tests` subfolder. The `dojoConfig` package should look something like the following:

```
Var basePath = location.pathname.replace(/\/[^\/]*$/, '');
var dojoConfig = {
  async: true,
  packages: [
    {
      name: 'app',
      location: basePath + '/js/'
    }, {
      name: 'spec',
      location: basePath + "/tests/"
    }
  ]
};
```

Since many of your custom widgets will require other modules, including other custom modules, to load, you need to keep the path for your custom modules the same as in your application.

In the body of the HTML document, add a script to call the specs through Dojo's `require()` function. The code will load the specs out of the tests subfolder through AMD. When the modules have loaded and the `dojo/ready` function is called, we will load Jasmine's `HtmlReporter` and execute the Jasmine testing:

```
<body>
<script type="text/javascript">
  require([
    "dojo/ready",
    "spec/Widget1",
    "spec/Widget2",
    "spec/Widget3"
  ], function (ready) {

    ready(function () {
      // Set up the HTML reporter - this is responsible for
      // aggregating the results reported by Jasmine as the
      // tests and suites are executed.
      jasmine.getEnv().addReporter(
        new jasmine.HtmlReporter()
      );
      // Run all the loaded test specs.
```

```
        jasmine.getEnv().execute();
      });
    });
  </script>
  </body>
```

When the tests are loaded and run, they will be rendered on this HTML page in the browser.

In the specs

Normally, Jasmine specs can either be run as a script or enclosed in a self-running JavaScript function. However, to work with Dojo, they need to be enclosed in module definitions. Unlike the custom modules we've made in the past, we won't use the dojo/_base/declare module to make a custom module. We'll simply use the define() statement to load the necessary modules, and run the test inside it.

For each suite, use the define() statement to help load the tests. You could also load any other ArcGIS JavaScript API modules or widgets you might need to help test the module. The code for the Widget1 spec, as requested by the main app, might look like the following:

```
define(["app/Widget1", "esri/geometry/Point"],
  function (Widget1, Point) {


});
```

Start writing Jasmine specs in the define statement. When the suite modules are loaded, they will automatically run any tests inside them and display the results on the SpecRunner page in the browser. You can use beforeEach to load a fresh module for testing, and afterEach to destroy it, if necessary. You can perform any custom ArcGIS JavaScript API-related tasks within the suites and specs:

```
define(["app/Widget1", "esri/geometry/Point"],
  function (Widget1, Point) {

  describe("My widget does some incredible tasks.", function () {
    var widget, startingPoint;

    // before each test, make a new widget from a staring point.
    beforeEach(function () {
      startingPoint = new Point([-43.5, 87.3]);
      widget = new Widget({start: startingPoint});
    });

    // after each test, destroy the widget.
    afterEach(function () {
```

```
      widget.destroy();
    });

    it("starts at the starting point", function () {
      expect(widget.start.X).toEqual(startingPoint.X);
      expect(widget.start.Y).toEqual(startingPoint.Y);
    });
    //…
  });

});
```

Once you have your tests defined, you can view the SpecRunner page in the browser and check out the results. If you have set everything up, you should have tests running and displaying results. Have a look at the following screenshot for an example:

```
Jasmine 1.3.1 revision 1354556913                        finished in 1.417s

. . . . . . . . . . . . . . . . . . . . . . . . . . . . . . . . . . . . .
. . . . . . . . . . . . .

Passing 53 specs                                              No try/catch ☐

A suite
  contains spec with an expectation

A suite is just a function
  and so is a spec

The 'toBe' matcher compares with ===
  and has a positive case
  and can have a negative case

Included matchers:
  The 'toBe' matcher compares with ===

  The 'toEqual' matcher
    works for simple literals and variables
    should work for objects
  The 'toMatch' matcher is for regular expressions
  The 'toBeDefined' matcher compares against `undefined`
  The 'toBeUndefined' matcher compares against `undefined`
  The 'toBeNull' matcher compares against null
  The 'toBeTruthy' matcher is for boolean casting testing
  The 'toBeFalsy' matcher is for boolean casting testing
  The 'toContain' matcher is for finding an item in an Array
  The 'toBeLessThan' matcher is for mathematical comparisons
  The 'toBeGreaterThan' is for mathematical comparisons
  The 'toBeCloseTo' matcher is for precision math comparison
  The 'toThrow' matcher is for testing if a function throws an exception
```

Use case for Jasmine and the ArcGIS JavaScript API

Jasmine provides an easy-to-write testing environment for your JavaScript widgets. If you are new to unit testing and BDD, and you aren't comfortable with setting up the Node.js environment on top of your tests, then Jasmine is a good place to start. All the libraries you need can be downloaded from GitHub, and you can begin writing tests almost right away.

 Jasmine can be downloaded from its GitHub repository at `https://github.com/jasmine/jasmine`. Documentation for Jasmine can be found at `http://jasmine.github.io/`. If you are interested in more reading material on this framework, you can read *Jasmine JavaScript Testing, Second Edition* by Paulo Ragonha, or *Jasmine Cookbook* by Munish Sethi.

Our application

For our application, we're going to set up unit testing for our Y2K map app using Intern. Unlike our other applications, which only required a browser, we're going to use Node.js for this setup. If we were following real TDD/BDD practices, we would have written these tests while we were developing the application. However, this will give you practice writing tests for any legacy applications you've previously written.

For this sample, we'll follow the example provided by David Spriggs and Tom Wayson through GitHub (`https://github.com/DavidSpriggs/intern-tutorial-esri-jsapi`). We'll make some modifications to work with more recent updates in both the ArcGIS JavaScript API and in the modules associated with the tests.

Adding testing files

In our Y2K application, we're going to add several files to work with the required Node.js modules. We'll add a `package.json` file, which tells the **Node Package Manager** (**NPM**) what modules we need and what version numbers we require. Next, we'll add a `Grunt.js` file, since we'll use Grunt to load the ArcGIS JavaScript API locally. We'll also add a `bower.json` file for Bower to load the correct `Dojo` libraries locally, and a `.bowerc` file to configure Bower when we run it. Finally, we're going to add an `intern.js` file in a `tests` folder to configure the Intern tests.

Our new folder structure should look like the following:

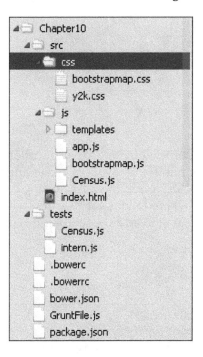

The package.json file

In our `package.json` file, we'll add the file dependencies required for our Node.js modules. When specifying version numbers for some of the modules, you'll note that some have a tilde (~) in front while others have a caret symbol(^). A tilde matches both major and minor version numbers (the first and second numbers in the three-number version code), while the caret will pull the most recent matching major version number. Copy the following into your `package.json` file:

```
{
  "name": "intern-tutorial-esri-jsapi",
  "repository": {
    "type": "git",
    "url": "https://github.com/DavidSpriggs/intern-tutorial-esri-
    jsapi.git"
  },
  "version": "0.2.0",
  "devDependencies": {
    "dojo": "^1.10",
    "esrislurp": "^1.1.0",
```

```
    "grunt": "^0.4",
    "grunt-contrib-watch": "~0",
    "grunt-esri-slurp": "^1.4.7",
    "intern": "^2.0.3",
    "selenium-server": "2.38.0"
  }
}
```

The Grunt setup

Grunt is a popular task runner used in Node.js applications. It's used to automate code building steps, such as minifying JavaScript, creating CSS files from CSS preprocessors like LESS or SASS, or, in this case, testing code. Grunt reads a Grunt.js file, which tells it where to look for files and what to do with them. In our Grunt.js file, we're going to add the following code:

```
module.exports = function(grunt) {
    grunt.initConfig({
        intern: {
            dev: {
                options: {
                    runType: 'runner',
                    config: 'tests/intern'
                }
            }
        },
        esri_slurp: {
          options: {
            version: '3.13'
          },
            dev: {
                options: {
                    beautify: false
                },
          dest: 'esri'
            }
        },
          esri_slurp_modules:{
          options: {
          version: '3.13',
          src: './',
          dest: './modules'
            }
        },
```

```
        watch: {
            all: {
                options: { livereload: true },
                files: ['src/js/*.js']
            }
        }
    });

    // Loading using a local copy
    grunt.loadNpmTasks('intern');
    grunt.loadNpmTasks('grunt-contrib-watch');
    grunt.loadNpmTasks('grunt-esri-slurp');

    // download Esri JSAPI
    grunt.registerTask('slurp', ['esri_slurp']);
    grunt.registerTask('create_modules', ['esri_slurp_modules']);

    // Register a test task
    grunt.registerTask('test', ['intern']);

    // By default we just test
    grunt.registerTask('default', ['test']);
};
```

In the preceding code, Node.js expects that we will pass our module definition to the `module.exports` variable. We tell Grunt where to find our `intern` module for testing in `grunt.initConfig`, and `esri_slurp` for downloading the ArcGIS JavaScript API. We also tell it to watch the `src/js/` folder for any changes to JavaScript files, which will trigger the tests to run again.

Once the configuration has loaded, we tell Grunt to load the modules for `intern`, `grunt-contrib-watch`, and `grunt-esri-slurp` to modules from `npm`. The tasks are then registered, and everything is set up to run the tests.

Setting up Bower

Bower is a Node.js module used to download all the required JavaScript libraries for a project automatically. We will use Bower to download the libraries and Dojo modules necessary to run our tests. Most common libraries and frameworks can be downloaded locally through Bower. Our exception in this case is the ArcGIS JavaScript API, which will currently be handled through `esri-slurp`. We need to fill out a `bower.json` file to tell Bower what libraries we need, and what library versions they should be as well.

Within our `bower.json` file, we need to create a JSON object to list our project name, version number, dependencies, and any development dependencies. We can name it whatever we want, so we'll name it `y2k-map-app`. We'll give it a version of `0.0.1` for testing, and we'll be sure to increment it whenever dependencies or library version numbers need to be updated. We have no development dependencies in this case, but we do need to list the parts of the Dojo framework, as well as `D3.js`, in this application.

We not only need to list the libraries and modules required for the tests, but the version numbers as well. We can refer to the ArcGIS JavaScript API *What's New* page at `https://developers.arcgis.com/javascript/jshelp/whats_new.html` to find out. According to the page, we need version `1.10.4` of the Dojo framework, including `dojo`, `dijit`, `dojox`, and `util` modules from `dojo/util`. We also need to add the `dgrid` modules with version `0.3.16`, `put-selector` version `0.3.6`, and `xstyle` version `0.1.3`. Referring back to the `dojoConfig` packages from our `index.html` page, we see that `d3.js` is using version `3.5.6`. We'll list the dependencies within a JSON object, with the version numbers preceded by a # sign. The contents of our `bower.json` file should be the following:

```
{
    "name": "y2k-map-app",
    "version": "0.0.1",
    "dependencies": {
        "dijit": "#1.10.4",
        "dojo": "#1.10.4",
        "dojox": "#1.10.4",
        "util": "dojo-util#1.10.4",
        "dgrid": "#0.3.16",
        "put-selector": "#0.3.6",
        "xstyle": "#0.1.3",
        "d3": "#3.5.6"
    },
    "devDependencies": {}
}
```

For the `.bowerc` file, we're going to add a simple configuration object. In the configuration object JSON, we're going to add a directories field and assign a value of `..` This will tell Bower to load the files requested in this folder. The contents of the `.bowerc` file should look like the following:

```
{"directories": "."}
```

Intern.js configuration

Now we need to set up our `Intern.js` configuration file in the `tests` folder to run our tests. We'll start by adding some default code in the page. More information on this configuration file can be found at `https://github.com/theintern/intern/wiki/Configuring-Intern`. Start by copying the following content:

```
// Learn more about configuring this file at
  <https://github.com/theintern/intern/wiki/Configuring-Intern>.
// These default settings work OK for most people. The options
  that *must* be changed below are the
// packages, suites, excludeInstrumentation, and (if you want
  functional tests) functionalSuites.
define({
  // The port on which the instrumenting proxy will listen
  proxyPort: 9000,

  // A fully qualified URL to the Intern proxy
  proxyUrl: 'http://localhost:9000/',

  // Default desired capabilities for all environments. Individual
  capabilities can be overridden by any of the
  // specified browser environments in the `environments` array
  below as well. See
  // https://code.google.com/p/selenium/wiki/DesiredCapabilities
  for standard Selenium capabilities and
  // https://saucelabs.com/docs/additional-config#desired-
  capabilities for Sauce Labs capabilities.
  // Note that the `build` capability will be filled in with the
  current commit ID from the Travis CI environment
  // automatically
  capabilities: {
    'selenium-version': '2.38.0'
  },

  // Browsers to run integration testing against. Note that
  version numbers must be strings if used with Sauce
  // OnDemand. Options that will be permutated are browserName,
  version, platform, and platformVersion; any other
  // capabilities options specified for an environment will be
  copied as-is
  environments: [{
    browserName: 'internet explorer',
    version: '10',
    platform: 'Windows 8'
  }, {
```

```
    browserName: 'internet explorer',
    version: '9',
    platform: 'Windows 7'
}, {
    browserName: 'firefox',
    version: '23',
    platform: ['Linux', 'Windows 7']
}, {
    browserName: 'firefox',
    version: '21',
    platform: 'Mac 10.6'
}, {
    browserName: 'chrome',
    platform: ['Linux', 'Mac 10.8', 'Windows 7']
}, {
    browserName: 'safari',
    version: '6',
    platform: 'Mac 10.8'
}],

// Maximum number of simultaneous integration tests that should
be executed on the remote WebDriver service
maxConcurrency: 3,

// Name of the tunnel class to use for WebDriver tests
tunnel: 'SauceLabsTunnel',

// Connection information for the remote WebDriver service. If
using Sauce Labs, keep your username and password
// in the SAUCE_USERNAME and SAUCE_ACCESS_KEY environment
variables unless you are sure you will NEVER be
// publishing this configuration file somewhere
webdriver: {
    host: 'localhost',
    port: 4444
},

// The desired AMD loader to use when running unit tests
(client.html/client.js). Omit to use the default Dojo
// loader
useLoader: {
    'host-node': 'dojo/dojo',
    'host-browser': 'node_modules/dojo/dojo.js'
},
```

```
// Configuration options for the module loader; any AMD
configuration options supported by the Dojo loader can be
// used here
loader: {
  // Packages that should be registered with the loader in each
testing environment
  packages:[]
},

// Non-functional test suite(s) to run in each browser
suites: [],

// A regular expression matching URLs to files that should not
be included in code coverage analysis
excludeInstrumentation: /^tests\//
});
```

Inside the `loader.packages` list, we'll add the folder locations of all the files we expect to have. These will include both the files we have now, plus the files we expect to download through Bower. Note that our `app` folder is referenced through the folder where our Census widget is stored. Also, we're loading all the ESRI and Dojo files, as well as `D3.js` for the graphics. Your packages portion should look like the following:

```
packages: [{
  name: 'tests',
  location: 'tests'
}, {
  name: 'app',
  location: 'src/js'
}, {
  name: 'gis',
  location: 'gis'
}, {
  name: 'esri',
  location: 'esri'
}, {
  name: 'dgrid',
  location: 'dgrid'
}, {
  name: 'put-selector',
  location: 'put-selector'
}, {
  name: 'xstyle',
  location: 'xstyle'
}, {
```

```
    name: 'dojo',
    location: 'dojo'
}, {
    name: 'dojox',
    location: 'dojox'
}, {
    name: 'dijit',
    location: 'dijit'
}, {
    name: 'd3',
    location: 'd3'
}
],
...
```

We need to add the names of our test suites to the suites list. For now, we'll add the following:

```
...
// Non-functional test suite(s) to run in each browser
    suites: [
    'tests/Census'
],
...
```

We should have everything we need in this file to run some successful tests on the `Census` widget. As we add more modules, we can expand this point with more tests.

We should be ready at this point to load the files we need through Node.js. If you haven't installed Node.js or any of the other node dependencies for this application, please follow the next section. If you already have, you can still review the next section, or skip it if you're already familiar with Grunt and Bower.

If you haven't installed Node.js

In this section, we'll go through the necessary steps to set up Node.js and the necessary dependencies. In these examples, we'll be working with NPM, Grunt, Grunt-CLI, and Bower. If you already have these set up in your development environment, you can skip to the next section.

If you haven't installed Node.js, you can download a copy from `http://nodejs.org`. On the home page, you'll find a link to the appropriate installer for your computer. Follow the directions in the installation wizard to install the software.

Node.js works through a command-line interface. All commands will be typed in either through a Linux or OSX terminal, or through the Windows command prompt (cmd.exe). Some portions may need you to have root or administrative access. Also, if you're using a Windows machine, the PATH variable may need to be updated to include both the node.js folder and its node_modules subfolder, where other commands and utilities will be loaded.

Next, if you have not installed the grunt, grunt-cli, and bower modules, we'll install them using npm. NPM comes built into Node.js and provides access to a multitude of plugins, libraries, and applications to help you generate applications faster. From your command prompt (or terminal), type the following to download the modules we need:

```
npm install -g grunt
npm install -g grunt-cli
npm install -g bower
```

The -g flag tells npm to install the modules globally, so that they can be accessed anywhere that Node.js and npm can be called. Without the -g flag, the module will be loaded in a node_modules subfolder of your current directory.

Loading all the libraries locally

Now it is time to load all of the Node.js applications necessary for testing. We're going to do the following:

1. Install all of our modules from npm.

2. Load most of the external JavaScript libraries locally using Bower.

3. Download the ArcGIS JavaScript API using Grunt Slurp.

We'll start by loading all the modules from npm. You will need to use a command line tool like Windows command prompt (cmd.exe) or terminal for Mac or Linux to run the code. Our package.json file defines the modules we need to run. We can install the modules by typing the following in the command line:

```
npm install
```

If we need to test for features in older browsers, such as Internet Explorer 8 or earlier, we will need to install an Intern extension called intern-geezer. It allows tests to work on older browsers that may not have the features used in standard Intern. If you need the older support, enter the following into the command line:

```
npm install intern-geezer
```

To load our JavaScript libraries locally, we can call on Bower to install the files. Our `bower.json` file tells it what to load, and our `.bowerc` file tells it where to load it. Enter the following into the command line:

bower install

Now, we need to load the ArcGIS JavaScript library for local testing. Since we defined everything for this task in our `Grunt.js` file, we can simply enter the following into the command line to download the files and start up our testing:

grunt slurp

Now our project should be up and running. We can now focus on writing tests for our application.

Writing our tests

We will be writing tests using the same `dojo` module definition style we used with creating objects, and similar to the Jasmine tests we talked about earlier. However, since Intern is designed for AMD, we can load the intern items within the module `define()` statements. In our tests folder, we'll create a file called `Census.js`, and let that contain our test module definition.

We will start by stubbing out our `define()` statement. We will load both Intern and Chai to run our tests. With Chai, we'll try using the `assert` module, where all test statements start with `assert`. We'll also load our `Census` widget and all the modules necessary to get it working, like a map. Since Intern typically runs without requiring HTML DOM, but our maps and `Census` widget require it, we'll add some modules to add elements to the DOM:

```
define([
    'intern!object',
    'intern/chai!assert',
    'app/Census',
    'esri/map',
    'dojo/dom-construct',
    'dojo/_base/window'
], function(registerSuite, assert, Census, Map, domConstruct, win) {
    //...
});
```

Within the module definition, we'll add the variables for our map and `Census` widget, and then register our suite with Intern using `registerSuite()`. In the `registerSuite()` function, we'll pass an object containing the test name `Census Widget`. We'll also add two methods, `setup()` and `teardown()`, to call before and after all the tests have run:

```
...
var map, census;

registerSuite({
  name: 'Census Widget',
  // before the suite starts
  setup:function () {},
  // after all the tests have run
  teardown: function () {},
});
...
```

In the setup function, we'll need to create the DOM, the map to attach to the DOM, and the `Census` widget to attach to the map. We'll use the `dojo/dom-construct` and the `dojo/_base/window` modules to create the DOM for both. Once the DOM elements have been created, we can initialize our map and census dijit. In our `teardown` function, we'll destroy the map so that it doesn't take up valuable memory space. The code should look like the following:

```
...
setup: function () {
  // create a map div in the body, load esri css, and create the
  map for our tests
  domConstruct.place('<link rel="stylesheet" type="text/css"
  href="//js.arcgis.com/3.13/dijit/themes/claro/claro.css">',
  win.doc.getElementsByTagName("head")[0], 'last');

  domConstruct.place('<link rel="stylesheet" type="text/css"
  href="//js.arcgis.com/3.13/esri/css/esri.css">',
  win.doc.getElementsByTagName("head")[0], 'last');

  domConstruct.place('<div id="map"
  style="width:300px;height:200px;" class="claro"><div id="census-
  widget"></div></div>', win.body(), 'only');
```

```
map = new Map("map", {
  basemap: "topo",
  center: [-122.45, 37.75],
  zoom: 13,
  sliderStyle: "small"
});

census = new Census({
  map: map,
  mapService:
  "http://sampleserver6.arcgisonline.com/arcgis/rest/
  services/Census/MapServer/"
}, "census-widget");
},
...
teardown: function () {
  map.destroy();
},
...
```

Now that the map and the Census widget have been initialized, we can start adding tests. I'll walk you through three relatively straightforward tests. It is up to you to extend them further. In this exercise, we'll test the Census dijit for the following:

- It should have all its working parts
- It should return the correct query results when we make a known request
- Parts of the charting widget should work as expected

When writing the first test, we'll give it a label of Test Census widget creation. For the corresponding function, we'll start by testing if the baseClass for the dijit is as expected. We'll use assert.strictEqual() to test whether the value is correct. Note that assert.strictEqual() takes three arguments, two values to compare, and a string description of the test. That test should look like the following:

```
...
'Test Census widget creation': function() {
  assert.strictEqual(
    census.baseClass,
    "y2k-census",
    "census.baseClass should return a string 'y2k-census'"
  );
},
...
```

Our second test may seem a little tricky, since it's an asynchronous test. However, Intern is designed to handle asynchronous testing as well. If necessary, we can tweak Intern's timeout before it considers the task failed, but for now we will just load the test.

In the query test, we'll send a query to the map service with all the states, and ask for a list of the states. Using a `then()` statement after the query, we can grab the results and test those using `assert.strictEqual()`. The test should look like the following:

```
...
'Test Census query for dropdown data': function () {

    census.queryShapeAndData({
    url:
    "http://sampleserver6.arcgisonline.com/arcgis/rest/
    services/Census/MapServer/3",
    fields: ["STATE_NAME", "STATE_FIPS"],
    where: "1=1",
    }).then(function (featureSet) {
    assert.strictEqual(
        featureSet.features.length,
        51,
        "There should be 51 states returned, including the District
        of Columbia"
    );
    });
},
...
```

Finally, for the graphing functions, we'll write tests to look at the functions we use to translate feature attribute data into a format that D3 can use. In this example, we'll pass attribute data through the `census.ethnicData()` method and test the output for expected values. Like other testing libraries, we can add more than one test within this specification. The test should look like the following:

```
'Test Census Graphing Attributes': function () {

  var ethnicAttributes = {
    WHITE: 10,
    BLACK: 20,
    AMERI_ES: 12,
    ASIAN: 11,
    HAWN_PI: 4,
    HISPANIC: 23,
    OTHER: 7,
```

```
    MULT_RACE: 17
  };

  var data = census.ethnicData(ethnicAttributes);

  assert.strictEqual(
    data.length,
    8,
    "The translation from graphic attributes to d3-based data
    should have 8 attributes in the ethnicData function"
  );
  assert.strictEqual(
    data[4].name,
    "Hawaiian / Pacific Islander",
    "The item in index 4 should have a name of Hawaiian / Pacific
    Islander data"
  );
  assert.strictEqual(
    data[5].population,
    23,
    "Out of the Hispanic column, the data index of 5 should have a
    population of 23."
  );
}
```

Checking the results

In order to check our tests, we need to load our file in the browser. Since we installed Intern as a node module, we'll need to view its test-running page. The file is located under `node_modules/intern/client.html`. You will need to specify where the tests are loaded in the browser by giving a query parameter of `config=tests/intern`. The following URL is an example you might view in a browser (it may be different depending on your setup): `http://localhost/MasteringArcGIS/Chapter10/node_modules/intern/client.html?config=tests/intern`.

When you correctly view the page, you'll be greeted with the Intern icon and a pass/fail report. Failing tests will be highlighted in red, while passing tests will be highlighted in green. You should see something like the following image:

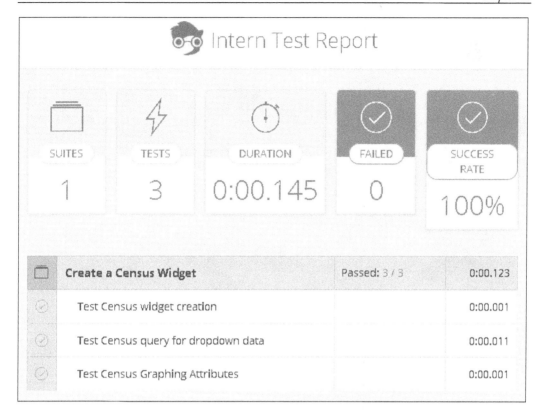

Summary

In this chapter, we have reviewed the reasons why we want to set up tests for our web applications. We have looked over different testing methodologies, development practices, and JavaScript libraries useful for testing our applications. Using Intern, we have set up tests for our own application.

In the next chapter, we'll look into the future of web applications for ArcGIS Server, checking into ArcGIS Online.

11
The Future of ArcGIS Development

What does the future hold for those who want to develop map applications for ArcGIS Server? After all, the direction of the Internet changes with each new technology, and what is popular today might not be popular tomorrow. The Flex and Silverlight APIs were quite popular at one point, and now they're only being maintained. What does the future hold for ArcGIS Server development with JavaScript?

If the latest releases from ESRI are any indication, JavaScript is still the development language of choice for the web. However, instead of letting users build applications from scratch, the ArcGIS platform has released new features that can jump start application development. Two of them, ArcGIS Online and Web AppBuilder, are based on the ArcGIS JavaScript API. We'll take a look at these technologies, and develop an application using a map created with ArcGIS Online.

In this chapter, we'll cover the following topics:

* How to use ArcGIS Online to create web maps
* What Web AppBuilder is and how it relates to ArcGIS Online
* How to use an ArcGIS Online web map in your ArcGIS JavaScript API application

ArcGIS Online

ArcGIS Online is ESRI's online tool for creating web maps in the cloud. ArcGIS Online provides a way for users to create web maps, mix and match map services and other data sources, modify popups, and share the resulting web maps, either with the general public or within their organizations. This is a screenshot of the ArcGIS Online page where you can add data from different sources into custom maps:

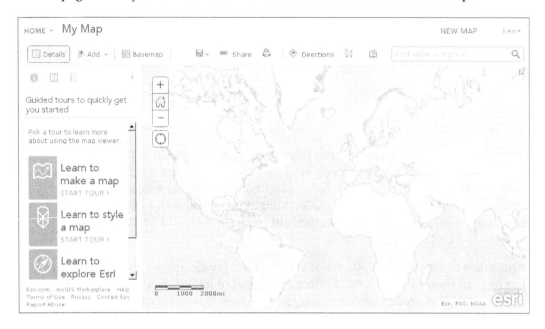

Besides referencing data through web maps, ArcGIS Online subscriptions allow you to upload and store data. Tiled map services can be created with tiled map packages generated through ArcGIS Server. Feature services can be created to store, update, and retrieve feature data. You can also use other ArcGIS Server extensions, such as geocoding and routing through ArcGIS Online. This means that you don't have to use your own hardware to host your data. You can maintain your feature data and tiled maps in the ArcGIS Online cloud.

Now, I'm sure you're wondering how ESRI can afford to offer these features. The more advanced services cost credits, which can be purchased through ESRI. Each simple transaction can cost a fraction of a credit, but repeated use can add up. Tiled and feature services also cost credits, and are charged by gigabytes of storage per month. Depending on the size of your organization, you get an allotted number of credits per year, with the option to purchase more. However, if you have a developer account, ArcGIS Online currently offers 50 credits a month for testing your applications.

ArcGIS Portal

If you work for an enterprise organization that doesn't want its data hosted through a public platform such as ArcGIS Online, there is a private version as well. ArcGIS Portal is an organization-centered version of ArcGIS Online that can work within a private or secured network. Logins can be verified through a number of common secure protocols such as SAML and oAuth2.

ArcGIS Portal has many of the same features as ArcGIS Online. You can create and share web maps. You can upload, catalog, and search for data sources in your organization and, using administrator access, you can set user roles and permissions. The primary difference is that the data you create and share is kept within the company network. What happens in ArcGIS Portal stays in ArcGIS Portal.

Web AppBuilder

Developers have made the same map tools over and over for different sites. Almost all generic web mapping applications have a drawing tool, a measuring tool, a print tool, a legend, and so on. The developers at ESRI China have implemented an idea to create an application that builds JavaScript web mapping apps that use ArcGIS Online web maps. They call the tool **Web AppBuilder**, that is shown in the following screenshot:

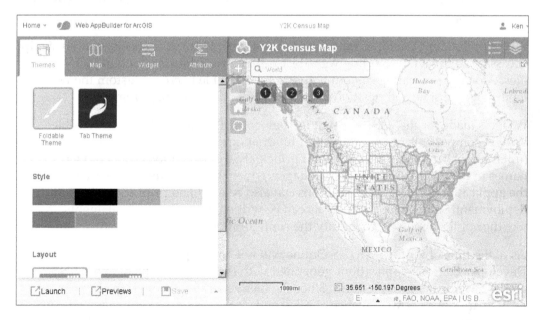

Web AppBuilder creates full-page web mapping applications using the ArcGIS JavaScript API. The applications use responsive design elements and fit any screen. Publishers have a choice of color scheme, title text, logos, and an assortment of widgets. Web AppBuilder applications are published and hosted through ArcGIS Online, so you don't need your own hosting environment to create a web map application.

You can create generic Web Appbuilder applications straight from ArcGIS Online. Using an existing AGOL webmap, you can create an application, add several default widgets, such as printing, measuring, and queries. You can define titles, color schemes (from a list), and parameters for each of the widgets you use.

Developer Edition

One issue with the default Web AppBuilder is that there is no room for custom development. If you don't like the way the **Query** widget searches for results, you don't have anything better. ESRI recognized this, and released a developer edition for custom development.

The Web AppBuilder Developer Edition is a Node.js application that lets you generate web mapping applications using default widgets and any custom widgets you create. Widgets are based on Dojo's `dijit` system, and are easily added to the system. Web AppBuilder and the applications created with it are tied to an ArcGIS Online account.

Web AppBuilder Developer Edition generates the HTML, style sheets, and JavaScript necessary to run a freestanding website. Configuration data is stored in JSON files. Along with creating custom widgets, it is possible to create custom themes, modifying the style and position of elements to match your requirements.

Widgets made for the Web AppBuilder Developer Edition are highly extensible. Widget components can be written for international markets using **i18n** (internationalization) and localization principals. Configurations can store label names in the widget in multiple languages, which will be loaded correctly with the application. Widget configurations can also be modified prior to application deployment. Settings for the widgets can be written using dijit form elements, much like the application itself, to modify the configuration.

As of the time of writing, ArcGIS Online will not host applications created with the Developer Edition. Applications generated with the Developer Edition are hosted on the developer's own platform. Since the files generated through the Developer Edition are static HTML, JavaScript, CSS, and image files, they can be hosted with little or no server-side processing.

For more information regarding Web AppBuilder, refer to the ESRI website at `http://doc.arcgis.com/en/web-appbuilder/`. For more information about Web AppBuilder Developer Edition, including download instructions and tutorials, visit `https://developers.arcgis.com/web-appbuilder/`.

More Node.js

Speaking of Node.js, there are a number of tools for the ArcGIS JavaScript API that rely on Node.js. The ArcGIS API for JavaScript Web Optimizer is a Node.js project hosted by ESRI, which packages your custom modules along with a minimum set of ESRI and Dojo modules for a simplified build process. You can host through ESRI's **content delivery network (CDN)**, or download and host it yourself.

If you decide to focus more on the Web AppBuilder Developer Edition, there's a Node.js project for that, too. The `generator-esri-appbuilder-js` project generates a template Web AppBuilder widget with all the necessary files. It asks you for basic information, such as the widget name and author information, as well as what kinds of files the widget will require. The project is based on Yeoman, a Node.js template-generating tool.

 For more information on the ArcGIS API for JavaScript Web Optimizer, visit `https://developers.arcgis.com/javascript/jshelp/inside_web_optimizer.html`. For the `generator-esri-appbuilder-js` project, you can find it on NPM at `https://www.npmjs.com/package/generator-esri-appbuilder-js`, or view the source code at `https://github.com/Esri/generator-esri-appbuilder-js`.

Our application

Now that we're up to speed on the latest and greatest from ESRI, let's return to our story. The Y2K society called and they love the work we've done with our application. They can't wait to share all that information with the general public.

The interesting thing is that the Y2K society has purchased a few ArcGIS Online licenses. Now they want you to store the web map data in that format, so they can manage and add new data about the year 2000 as they come across it.

Our goal is to create an ArcGIS Online webmap for the Y2K society and add it to our existing application. We will add the map layers, popup templates, and charts into the webmap. We'll then load the webmap into our existing application, cut out unnecessary features, and make the rest of the custom features work with the webmap.

Creating a webmap

We need to create a webmap to use in our application. If you haven't already set up a developer account for ArcGIS Online, go to `https://developers.arcgis.com/en/sign-up/` to sign up for a developer account. Once you've signed in, click on **Map** in the top menu. It will load an empty map much like the image at the beginning of the chapter.

We can start by saving a copy of our map. This will allow us to name the map, write descriptions, search terms, and so on. In the toolbar toward the top, click on the **Save** dropdown and select **Save**. This will create a dialog popup that will let us enter the name of our map, write a summary, and add tags that make it easier to search for our map. We can put the map in our base folder, or make a new one for the application. Add `Y2K Census Map` to the title, and give a brief description. It should look similar to the following screenshot:

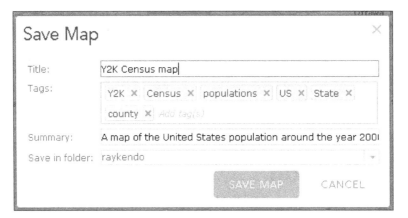

Once you've clicked **SAVE MAP**, the map should have a new name. If you look in the URL, you'll also notice some strange characters. Your webmap has now been assigned a new webmap ID (`?webmap=XXXXXXXXXXXXXXXXX`), which is a unique string that ArcGIS Online uses to reference your map. Keep this webmap ID handy, because we'll use it later in our application.

Changing the basemap

One of the best features of using ArcGIS Online to host your webmap is all the basemap backgrounds that are available. From satellite and aerial photos to simple grayscale maps, there's a good background available for your data.

In previous versions of our application, we used the national geographic basemap to frame our data. We can do the same for this webmap. Click **Basemap** in the upper left-hand corner of the toolbar, and you'll be able to view a number of web map thumbnail images. Select the **National Geographic** basemap, and the map should update. The basemap selector should look like the following screenshot:

Adding the census map service

ArcGIS Online provides a number of ways to add layers to maps. Whether we have an ArcGIS Server map service or a simple CSV file, ArcGIS Online can accommodate a wide variety of data. To add layers to the map, click on the **Add** button in the upper left-hand corner and select from the drop-down choices. Some of the different ways to add layers to webmaps are discussed in the following sections.

Searching for layers

ArcGIS Online catalogs a wide variety of map layers for both public and organizational use. ArcGIS Online provides a search blank and a list of matching maps and layers to add to your webmap. If layers and services have been shared within your organization, you can search for keywords and tags to add that layer into your map. You can also add map services and layers that other people have marked as public data.

 If your web map contains layers that are restricted to either your organization, or to specific groups within your organization, anyone who views the webmap will be prompted to log in to prove they can view the layers. If login fails, the webmap may load, but the restricted data will not be loaded on the map.

Browsing for ESRI layers

ESRI has a large collection of its own map data available for public use. The data covers a wide range of topics and interests. You can filter by category, and click to add the map layer to your existing map. Currently, major categories include the following:

- Imagery
- Basemaps
- Historical maps
- Demographics and lifestyles
- Landscapes
- Earth observations
- Urban systems
- Transportation
- Boundaries and places

While you might be tempted to grab census data from this location, remember that this shows more current data. Since we're using the Year 2000 data in this case, we'll ignore this option. However, we'll remember this, because it could be useful in other projects.

Adding layers from the web

This option allows us to add from a number of web map services. While it's expected that we can add ArcGIS Server map services, some of the other choices may surprise you. Besides ArcGIS Server map and feature services, other layer types we can add include:

- WMS Service
- WMTS Service
- Tiled Service
- KML file
- GeoRSS file
- CSV file

Note that the CSV file refers to one available online, more than likely through another service. If you have a CSV file to upload, you would select the next option.

Adding layer from a file

Here, we get to upload a file and have ArcGIS Online render it for us. The dialog asks for a file to upload and, based on the file format, it will also ask for information to determine geographical locations. Acceptable files for adding geographical data include the following:

- A shape file with up to 1,000 features stored in a `.zip` folder
- A tab or comma delimited text or CSV file
- A **GPS Exchange Format (GPX)**, which is available from many GPS units

Adding map notes

This option allows you to draw custom points, lines, and polygons on the map as notes for other users or applications. You can select from a number of predefined feature types with unique renderers. Based on the note template you pick, you can add points, lines, and polygons with custom symbols.

Our application data source

For our application data source, we'll add a layer from the web. After you choose to add a layer from the web, select the ArcGIS web service from the drop-down list, and either type or paste the 2000 census map service URL in the blank. Leave the **Use as Basemap** checkbox unchecked. Finally, click **ADD LAYER** to add it to the map. You can see a screenshot of the prompt in the following image:

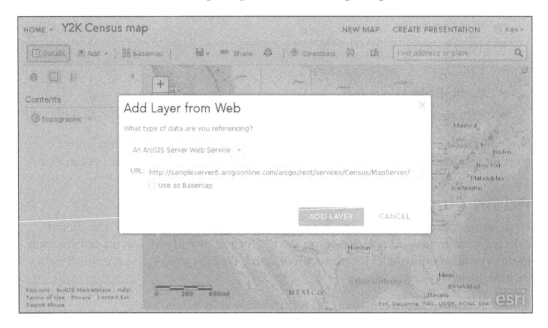

Styling

ArcGIS Online allows you to restyle many dynamic and feature layers, even if they weren't styled that way on the original map. You can add custom renderers to alter the color, transparency, and picture symbology of each layer. You can turn a simple map with no style into a custom map with unique value symbology, telling the story you want to tell.

For our web map, we see the simple black outlines of the states and counties on our map. We're going to change the style to make them look more appealing, and to blend them in with the National Geographic basemap background.

First, click on the census map service layer under **Contents** to expand the map layer. You'll see the different sublayers that make up the layer, such as **States**, **Counties**, **Block Groups**, and **Block Points**. By clicking on each one, you can see the symbols for each layer.

We'll start changing the symbology for the **States**. Click the little downward-pointing triangle to the right of **States** and select **Change Style**. The first menu will appear on the left, and let you select a label and symbol. Between the basemap and our popups, we'll have a good idea of what state we're in, so we'll leave the labels alone. Click on the **Options** button in the **Single Symbol** drawing under **Select a drawing style**.

The next prompt lets you change the symbol and the visibility range of the layer on the map. If the layers are already set with minimum and maximum visibilities, this will not make those more lenient, but it can make them stricter. Click on the **Symbol** link to get a color prompt. From here, you can set the color of the outline and the inside filling of the polygons. It should look something like the following screenshot:

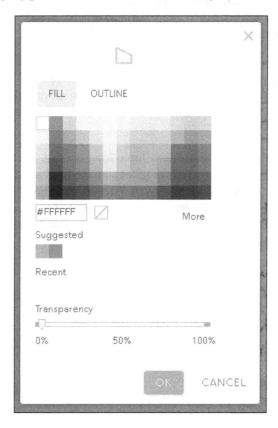

Change the color of the outline to a darker orange, and click **OK**. We'll leave a transparent fill in the states. Repeat the same processes with the **Counties** and **Block Groups**, picking a lighter shade of orange each time.

Creating popups

Now that we've adjusted our layers, let's look at the popups. ArcGIS Online lets you create and configure popups for map layers and feature services on your webmap. They provide a number of popup configurations, including a list of feature attributes and a **What You See Is What You Get** (**WYSIWYG**) rich text editor. Both allow some value formatting, such as thousand separators for numbers, or date formatting for dates. The WYSIWYG editor also allows you to use text colors and add links based on value attributes.

For our application, we're going to stick with displaying the list of attributes. Click on the **Options** button for the **States** (the little downward-pointing triangle), and select **Configure Popup**. In the next menu on the left-hand side, we can modify the title and the contents. Remember that, in any of the blanks, field names are surrounded by brackets and will be replaced with values.

Next to the **Display** dropdown, select **A list of field attributes**. This will create a long list of all the field attributes. We don't want to see all of them, or even most of them, so we'll configure which fields we see. Click on the **Configure Attributes** link under the list of fields.

You'll see a dialog with a list of fields available for viewing. Each field has a checkbox for visibility, a field name shown with brackets, and a field alias. Field aliases can be altered by clicking on them and typing in a different value. Unchecking a box here means that it won't appear in the popup. For numeric and date fields, extra formatting prompts appear to the right of the list when such a field is selected. Look at the following screenshot:

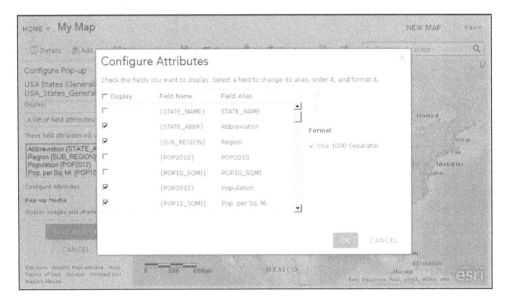

Since we see the state name in the title, we can uncheck everything but **Population for 2000**, **Square Miles**, and **Population per Square mile**. We'll rename the aliases to make them more appealing. Finally, we'll click **OK**. The list of attributes should reflect the three fields we've made visible at this point.

Adding charts

The popup configuration controls on ArcGIS Online also allow you to add images, charts, and graphs based on feature attributes. The images and graphs appear in the popups under the content we've displayed. These graphics are drawn using the `dojox/charting` libraries.

Going back to the **States** layer, we'll add the graphs for the ethnicities, genders, and ages in the states. In the lower left, under **Popup Media**, click on the **Add** button. A dropdown will appear, letting you choose the image or chart you would like to see. For our ethnic graph, we'll select a pie graph.

In the next prompt, we're given the parameters required to configure the pie chart. It requires a title, a caption, and selected fields to add to the chart. We're also given the option to normalize the data against another field in our data. If we were comparing relative population levels between areas, it would be beneficial to normalize against the entire population, but since we're making pie charts, and the populations of each ethnic group should add up to 100 percent, we'll skip normalization. You can see an example of the prompt in the following screenshot:

For the **State** popup, we'll give the graph a name of `Ethnic Graph`. We'll then add a caption to explain what the chart is all about. Finally, we'll make sure that the **WHITE, BLACK, AMERI_ES, ASIAN, HISPANIC, MULT_RACE, HAWN_PI**, and **OTHER** fields are checked. Note that, if you didn't reformat the names of the fields when configuring the popup field name attributes in the previous section, the attributes will appear just as you left them. You can edit the field aliases here as well. You don't necessarily have to make them visible in the popup to make them chartable (and format their aliases) here. Finally, click **OK**, and you should see your **Ethnic Chart** in the list of **Popup Media**.

We'll repeat the steps for adding the gender and age charts. For the gender pie chart, we'll repeat the preceding steps to add another pie chart. We'll label it `Gender Chart`, and make sure that the `MALES` and `FEMALES` fields are checked. For the age bar chart, we'll select **Bar Chart** from the first menu. In the second prompt, we'll title the chart `Age Chart` and add the `AGE_UNDER5, AGE_5_17, AGE_18_21, AGE_22_29, AGE_30_39, AGE_40_49, AGE_50_64`, and `AGE_65_UP` fields.

Once the charts have been configured, be sure to click on **SAVE POPUP** on the bottom of the popup menu. This will reload the map with the new popup configurations. Click on a state to test that the popups appear. You should see the fields we added, and **Ethnic Chart** in the **State** popup. Arrows are also added to the right and left of the charts so that the user can view the gender and age graphs as well. Hovering over the graphs with your mouse will show the labels and numbers behind the graphs. They should appear like the popups shown in the following pictures.

This screenshot shows the **Ethnic Graph**:

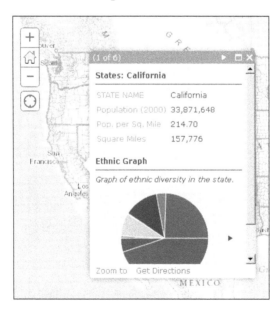

This screenshot shows the **Gender Chart**:

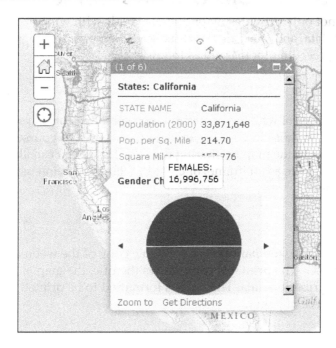

This screenshot shows the **Age Chart**:

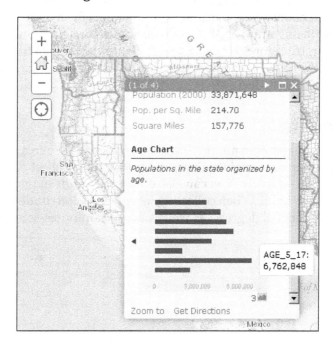

Uses for an ArcGIS Online webmap

ArcGIS Online not only provides maps, but also tools with which you can perform analysis, collect data, and share maps with others. Some of the other features available in the preceding toolbar are as follows.

Sharing

Sharing lets you control who has access to your webmap. You can share the map with the world, with those in your organization, or keep it to yourself. You can give other people access to the map by sharing the URL for it. You can also imbed the webmap in any website as an iframe, or create a web application using one of the web app templates.

Printing

The **Print** control in the toolbar opens a printable copy of the webmap in a new tab on your browser. The printable copy shows the title, the map in its current extent, and the attribution data. The page is formatted to be printed on an ANSI (A letter-sized) piece of paper.

Directions

By clicking on **Directions**, you'll get a Directions widget in the left-hand portion of the screen. The Directions widget lets you get directions from a starting point to a finishing point, with any number of stops in between. Directions and time can be calculated for a car, truck, or walking. You can also collect return trip data, as well as traffic data, on the route. Remember that directions and traffic routing cost credits to use.

Measurements

The measurement widget here mimics the measurement `dijit` in `esri/dijit/Measurement`. You can draw points, lines, and polygons to collect coordinates, lengths, and areas respectively. You can also change the units of measurement into metric or English units. When collecting latitude and longitude with points, it displays them in decimal degrees, as well as degrees-minutes-seconds.

Bookmarks

The **Bookmarks** tool allows you to save locations on the webmap. The bookmarks widget captures the current map extent and gives it a name. When creating multiple bookmarks of areas on the map, you can quickly navigate to them by clicking on the bookmark names. You can also edit the names and delete unwanted bookmarks. Bookmark data is saved along with other features of the webmap.

Developing against an ArcGIS Online webmap

If your organization is happy with the webmaps and tools offered by ArcGIS Online, you could potentially run your public facing GIS from the website and never develop another tool. Sometimes you need custom functionality. Sometimes you need an application that makes custom requests from other data sources when you click on a map. Maybe you need a complicated query page or a query builder to collect results. Whatever your specialized needs, you can bring the ArcGIS Online webmap to your custom app.

What does using an AGOL webmap do for your application? It cuts out many of the map configuration steps. All the tiled, dynamic, and feature layers we would normally hardcode in the application can now be mashed together in a webmap and loaded in our app. All the popups and `infoTemplate` that we had to assign through code can now be carried over from the we map. Whenever changes to the map service layers and popups are required, they can be made through ArcGIS Online, instead of changing the code and hoping that users aren't still running a cached copy of your configuration script. Configurations can be handled by someone else, giving you more time to code.

Creating the webmap

In order to bring in the webmap we created earlier, we need to update our `app.js` file. In the list of modules in the `app.js` file's `require()` method, we need to add a reference to the `esri/arcgis/utils` module. This will give us the tools we need to add the webmap to our application:

```
require([, "esri/arcgis/utils", ],function (, arcgisUtils, ) {

});
```

Typically, we would use the `arcgisUtils.createMap()` method to construct our new webmap. It requires parameters similar to the `esri/map` constructor, including a reference to the HTML element `id` where the map should go, and an `options` object. However, the `arcgisUtils.createMap()` method also requires an ArcGIS Online webmap `id`. You can get this `id` by opening your webmap in ArcGIS Online to edit. If you look at the URL, you'll see the search parameter `webmap=XXXXXXXXXXXXXX`, where those X's are replaced by alphanumeric characters. You can copy and paste the `id` numbers into your application to create a webmap, like so:

```
require([, "esri/arcgis/utils", ],function (, arcgisUtils, ) {
    arcgisUtils.createMap("450d4fb709294359ac8d03a3069e34d3",
        "mapdiv", {});
});
```

Options for the `arcgisUtils.createMap()` method include, but are not limited to the following:

- `bingMapKey`: The unique key string provided in your subscription to Bing Maps.
- `editable`: A `Boolean` value that you can use to disable any feature layer editing.
- `geometryServiceURL`: URL link to ArcGIS Server geometry service.
- `ignorePopups`: A `Boolean` that will disable all popups if true.
- `mapOptions`: The normal options you would pass to create a normal map.
- `usePopupManager`: A `Boolean` value that, when true, tells `InfoWindow` to retrieve all clicked features that have popup templates. When true, it also handles the popups visibility of map service sublayers whether the parent layer is visible or not. If false, popups on sublayers may or may not load if the parent layer is not turned on, and you will have to manually control popup visibility for complicated maps.

The `arcgisUtils.createMap()` method will return a `dojo/Deferred` promise, which will fire a success callback using `then()` once the webmap data has been downloaded and the map has been created. The callback will return a response object that contains the following:

- `map`: The `esri/map` object created
- `itemInfo`: An object containing:
    - `item`: An object that contains the map title, a snippet to describe it, and the extent in a nested array form

- ○ `itemInfo`: A JSON object containing data about the map services, renderers, popups, bookmarks, and other map data used to create the web map

- ○ `errors`: An array of errors encountered when loading data and creating the map services

- `clickEventHandler` and `clickEventListener`: When populated, these control the map click events that show popups. If you ever need to turn off popups, you can run the following:

```
// response object was returned by arcgisUtils.createMap()
response.clickEventHandler.remove();
```

- To restore the popups, run the following line of code (assuming that you've loaded the `dojo/on` module:

```
response.clickEventHandler = dojoOn(map, "click",
  response.clickEventListener);
```

> When creating the webmap, if you set the `usePopupManager` parameter to `true` (the default is `false`), the `clickEventHandler` and `clickEventListener` parameters will be null. In that case, popup visibility will be controlled through the `map.setInfoWindowOnClick(boolean)` method.

Since we incorporated the ESRI Bootstrap back in *Chapter 8, Styling Your Map*, we need to use the ESRI Bootstrap version of the `createMap()` constructor in the `esri/arcgis/utils` module. Since the map may take time to load, we'll move the census widget loading inside the deferred callback. The code should look as follows:

```
/* not used because we're using the boostrapMap version
var deferred =
arcgisUtils.createMap("450d4fb709294359ac8d03a3069e34d3", "map",
{}); */

var deferred =
BootstrapMap.createWebMap("450d4fb709294359ac8d03a3069e34d3",
"map", {});

deferred.then(function (response) {
  // map and census widget definition moved inside deferred
  callback
  var map = response.map;
```

```
var census = new Census({
  map: map,
  mapService:
  "http://sampleserver6.arcgisonline.com/arcgis/rest/services/
  Census/MapServer/"
}, "census-widget");

dojoOn(dom.byId("census-btn"), "click", lang.hitch(census,
census.show));
});
```

In our census widget

Now that we have defined many parts of our map in the webmap, we can remove a lot of the widget code that handled the same information. In the module reference of our census's `define` statement, we can remove references to the following modules:

- The **State**, **County**, **Block Group**, and **Block Point** templates
- `IdentifyTask` and `IdentifyParameter`
- `InfoTemplates`
- `D3/d3`

Remember to delete the module references as well as their corresponding variables.

In the body of our widget, we can remove any references to the `identifytask` or map clicking, or `infowindow` selection events, since that controlled the popups. We can also remove all the methods that created the charts on our popups. All we're left with is a widget that populates the drop-down menus for states, counties, and block groups, to let us see that data.

Don't feel bad about deleting so much code. One of the joys of a developer's life is to delete lines of code, since they are possible sources of errors.

Accessing webmap in our app

What would we do if we needed something from the ArcGIS Online webmap configuration? If we want that information, the ArcGIS JavaScript API gives us the tools to retrieve it. By using the `esri/arcgis/utils` and `esri/request` modules, we can retrieve and work with our web map data.

To access our web map configuration data, we need two pieces of information: the URL of the site hosting the data and the web map ID. The URL of the service hosting the webmap data can be found in the `arcgisUrl` property of the `esri/arcgis/utils` module. By default, it points to `http://www.arcgis.com/sharing/rest/content/items`, but if your webmaps are hosted in a private enterprise network in the ArcGIS portal, this URL string should be replaced with the appropriate URL for your portal service.

In our JavaScript code, we can combine the URL hosting the webmap data, along with the webmap ID, into a URL to access JSON data for our webmap. We can use the `esri/request` module to request the data from the URL and use it wherever we need in our application. A bit of sample code is shown in the following snippet:

```
require([…, "esri/arcgis/utils", "esri/request", …],
  function (…, arcgisUtils, esriRequest, …) {
    …
    // replace fake webmapid value with real webmap id
    var webmapid = "d2klfnml2kjfklj2nfkjeh2ekrj";
    var url = arcgisUtils.arcgisUrl + "/" + webmapid + "/data";
    esriRequest({
      url: url,
      handleAs: "json",
      content: {f: "json"}
    }).then(function (response) {
      console.log(response);
      // do something with the JSON data.
    });
    …
  });
```

If you look at the JSON results from our webmap, using either the map console or a JSON viewing plugin for the browser, you can see that the webmap data for our application includes map service layers, a background basemap, and even details about the configuration of our popups. There's a lot of potentially useful information in this response, should we need it.

Interacting with the map through code

In our previous version, we loaded `InfoTemplates` for each layer to define the format of the content displayed in the popups. Now that we have all that information defined through ArcGIS Online, we don't need to make the extra HTTP requests to load the HTML templates. You can click on the map and the popups will be ready to be used.

However, we run into a problem when we try to use our custom drop-down selectors to bring up the popups. In the older version, we would query the data, add an appropriate template, and insert it in the `InfoWindow`. Now, a user click pulls the appropriate data for us. But how can we click on the map without clicking on the map?

Using an undocumented trick, we can trigger a map-click event with our data. We could zoom to the feature extent on the map and trigger the map click event when it's done. This trick has been successfully deployed on a number of websites that use ArcGIS Online webmaps.

Look for the `_updateInfoWindowFromQuery()` method in the `Census.js` widget. It currently contains steps that set the map extent and fill in the contents of the map's `InfoWindow`. We'll keep the extent setting, but remove any references to `map.infoWindow`. Instead, inside the asynchronous callback function called after the map has changed extents, we'll add code to call the `map.onClick()` method to trigger the click event:

```
_updateInfoWindowFromQuery: function (results) {
  var resultExtent =
  graphicsUtils.graphicsExtent(results.features);
  this.map.setExtent(resultExtent).then(lang.hitch(this, function ()
  {
  // do something
  this.map.onClick();
  }));
},
```

The `map.onClick()` function requires an object with three properties: a graphic, a `mapPoint`, and a `screenPoint` that contains the coordinates of the point on the screen. We can use the first of our query results for the graphic that has been clicked. For a quick and dirty click location, we can use the center of the current extent, which we get from `resultExtent.getCenter()`. As for the corresponding `screenPoint` for our map point, the map has methods for converting geographic points to screen points and the other way around. We'll use the `map.toScreen()` method with the center point to get its coordinates on the screen. We'll then feed the results into the `map.onClick()` method, and test the application. The function should look like the following:

```
_updateInfoWindowFromQuery: function (results) {
  var resultExtent =
  graphicsUtils.graphicsExtent(results.features);
```

```
    this.map.setExtent(resultExtent).then(lang.hitch(this, function
    () {
    var center = resultExtent.getCenter();
    var centerScreen = this.map.toScreen(center);
    this.map.onClick({
        graphic: results.features[0],
        mapPoint: center,
        screenPoint: centerScreen
    });
    }));
},
```

We can test the code by loading our application and clicking the **Census** button. It should show our drop-down modals for the state, county, and block group data. When we select a state like California, it should query, zoom, and the popup should appear momentarily with the popups we configured earlier. Now, we have a fully functioning census application, while letting ArcGIS Online handle most of the work:

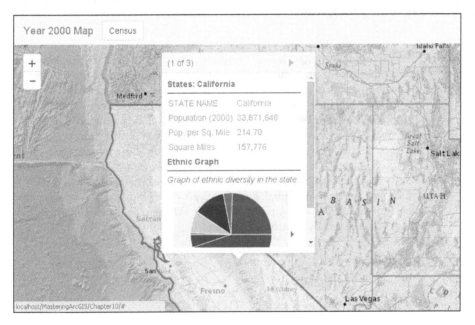

Summary

In this chapter, we have looked at the newest tools provided by ESRI for the ArcGIS platform: ArcGIS Online and Web AppBuilder. We have created a webmap using ArcGIS Online, adding layers, formatting them, and creating popups. We added the webmap to our own application.

Throughout this book, we've dived deep into the ArcGIS JavaScript API. We started by creating a simple webmap application, complete with click events and geographical data. You have learned how to use to tools provided by the API and the Dojo framework to build custom controls and modules that you could reuse from project to project. You then experimented with adding other JavaScript libraries, so that developers not so familiar with dojo could contribute. You learned how to edit geographical data through web applications. You also learned how to style your map applications for desktop and mobile use. You even learned how to unit test your code. Finally, you learned how ESRI is simplifying webmap development through ArcGIS Online and Web AppBuilder, bringing applications to the public faster. You have definitely come a long way.

 Where do you go from here? There are a number of great books, blogs, and other resources. You can find out about current trends in ArcGIS development through ESRI's forums at `http://geonet.esri.com`, or find support for the latest ArcGIS APIs at `http://developers.arcgis.com`. For blogs, you can check out Dave Bouwman's blog at `http://blog.davebouwman.com`, or Rene Rubalcava at `http://odoe.net`. For books, you may want learn about how to manage ArcGIS Server to serve your web maps with *Managing ArcGIS for Server* by Hussein Nasser. You may also want to check out *ArcGIS Web Development* by Rene Rubalcava, *Developing Web and Mobile ArcGIS Server Applications with JavaScript* by Eric Pimpler, or *Developing Mobile Web Applications* by Matthew Sheehan.

Index

Symbols

_FocusMixin module
adding 71
_Mixins
working with 70
_OnDijitClickMixin module
adding 71
_TemplatedMixin module
adding 70
_WidgetBase module
about 69
buildRendering 69
constructor 69
Custom setters are called 69
destroy 69
methods 69, 70
parameters mixed into widget instance 69
postCreate 69
postMixinProperties 69
startup 69

A

AccordionContainer 228
AccordionPane 229
all-in-one Editor dijit 59
Angular
MV* 213
overall results 216
AngularJS 213
Angular vocabulary
about 214
app controller 214
app directive 214
app service 214
API features
defining 2
app
about 284, 305
Census dijit modal, creating 235-239
ESRI-Bootstrap, adding 231, 232
HTML, bootstrapping 232
libraries, loading locally 293, 294
Node.js, setting up 292, 293
restyling 231, 234
results, checking 298
testing files, adding 284
tests, writing 294-297
**Application Programming
Interface (API) 29**
Aptana Studio 3
reference 3
ArcGIS API for JavaScript Web Optimizer
reference 305
ArcGIS compact build
about 245
modules 245
ArcGISDynamicMapServiceLayer 38
ArcGIS JavaScript API
about 1, 151
Deferreds 20
defining 17
events 17
promises 20
tasks 19
URL 1

About Packt Publishing

Packt, pronounced 'packed', published its first book, *Mastering phpMyAdmin for Effective MySQL Management*, in April 2004, and subsequently continued to specialize in publishing highly focused books on specific technologies and solutions.

Our books and publications share the experiences of your fellow IT professionals in adapting and customizing today's systems, applications, and frameworks. Our solution-based books give you the knowledge and power to customize the software and technologies you're using to get the job done. Packt books are more specific and less general than the IT books you have seen in the past. Our unique business model allows us to bring you more focused information, giving you more of what you need to know, and less of what you don't.

Packt is a modern yet unique publishing company that focuses on producing quality, cutting-edge books for communities of developers, administrators, and newbies alike. For more information, please visit our website at www.packtpub.com.

Writing for Packt

We welcome all inquiries from people who are interested in authoring. Book proposals should be sent to author@packtpub.com. If your book idea is still at an early stage and you would like to discuss it first before writing a formal book proposal, then please contact us; one of our commissioning editors will get in touch with you.

We're not just looking for published authors; if you have strong technical skills but no writing experience, our experienced editors can help you develop a writing career, or simply get some additional reward for your expertise.

ArcGIS for Desktop Cookbook

ISBN: 978-1-78355-950-3   Paperback: 372 pages

Over 60 hands-on recipes to help you become a more productive ArcGIS Desktop user

1. Learn how to use ArcGIS Desktop to create, edit, manage, display, analyze, and share geographic data.

2. Use common geo-processing tools to select and extract features.

3. A guide with example-based recipes to help you get a better and clearer understanding of ArcGIS Desktop.

Learning ArcGIS Geodatabases

ISBN: 978-1-78398-864-8   Paperback: 158 pages

An all-in-one start up kit to author, manage, and administer ArcGIS geodatabases

1. Covers the basics of building Geodatabases, using ArcGIS, from scratch.

2. Model the Geodatabase to an optimal state using the various optimization techniques.

3. Packed with real-world examples showcasing ArcGIS Geodatabase to build mapping applications in web, desktop, and mobile.

Please check **www.PacktPub.com** for information on our titles

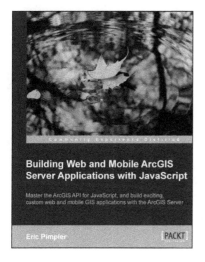

Building Web and Mobile ArcGIS Server Applications with JavaScript

ISBN: 978-1-84969-796-5   Paperback: 274 pages

Master the ArcGIS API for JavaScript, and build exciting, custom web and mobile GIS applications with the ArcGIS Server

1. Develop ArcGIS Server applications with JavaScript, both for traditional web browsers as well as the mobile platform.

2. Acquire in-demand GIS skills sought by many employers.

3. Step-by-step instructions, examples, and hands-on practice designed to help you learn the key features and design considerations for building custom ArcGIS Server applications.

Programming ArcGIS 10.1 with Python Cookbook

ISBN: 978-1-84969-444-5   Paperback: 304 pages

Over 75 recipes to help you automate geoprocessing tasks, create solutions, and solve problems for ArcGIS with Python

1. Learn how to create geoprocessing scripts with ArcPy.

2. Customize and modify ArcGIS with Python.

3. Create time-saving tools and scripts for ArcGIS.

Please check **www.PacktPub.com** for information on our titles